Water Quality - Science, Assessments and Policy

Edited by Kevin Summers

Published in London, United Kingdom

IntechOpen

Supporting open minds since 2005

Water Quality - Science, Assessments and Policy
http://dx.doi.org/10.5772/intechopen.77531
Edited by Kevin Summers

Contributors
Kate Mulvaney, Nathaniel Merrill, Marisa Mazzotta, Limin Ma, Changxu Han, Steve Paulsen, David Peck, Philip Kaufmann, Alan Herlihy, John Kiddon, Hugh Sullivan, Marguerite Pelletier, Linda Harwell, Walter Nelson, Mari Nord, Nayla Omer, Mary Kentula, Amanda Nahlik, Teresa Magee, Hongbo Du, Raghava R Kommalapati, Abdallah Zacharia, Anne H Outwater, Rob Van Deun, Subhani Bandara, Laura E. Carson

First published in London, United Kingdom, 2020 by IntechOpen
IntechOpen is the global imprint of INTECHOPEN LIMITED, registered in England and Wales, registration number: 11086078, 7th floor, 10 Lower Thames Street, London, EC3R 6AF, United Kingdom
Printed in Croatia

British Library Cataloguing-in-Publication Data
A catalogue record for this book is available from the British Library

Additional hard and PDF copies can be obtained from orders@intechopen.com

Water Quality - Science, Assessments and Policy
Edited by Kevin Summers
p. cm.
Print ISBN 978-1-78985-577-7
Online ISBN 978-1-78985-578-4
eBook (PDF) ISBN 978-1-83968-392-3

For EU product safety concerns:
IN TECH d.o.o., Prolaz Marije Krucifikse Kozulić 3, 51000 Rijeka, Croatia, info@intechopen.com or visit our website at intechopen.com.

Meet the editor

Kevin Summers is a Senior Research Ecologist at the Environmental Protection Agency's (EPA) Gulf Ecosystem Measurement and Modeling Division. At present, he is working with colleagues in the Sustainable and Healthy Communities Program to develop an index of community resilience to natural hazards, an index of human well-being that can be related to changes in ecosystem, social and economic services, and a community sustainability tool for communities with populations less than 40,000. He leads research efforts for indicator and indices development. Dr. Summers is a systems ecologist and began his career at the EPA in 1989 and has worked in various programs and capacities. This includes leading the National Coastal Assessment in collaboration with the Office of Water culminating in the award-winning National Coastal Condition Report series (four volumes between 2001 and 2012), which integrates water quality, sediment quality, habitat and biological data to assess the ecosystem condition of the estuaries of the United States. He was the acting National Program Director for Ecology for the EPA between 2004 and 2006. He has authored approximately 150 peer-reviewed journal articles, book chapters, and reports and has received many awards for technical accomplishments from the EPA and outside the agency. Dr. Summers holds a BA in Zoology and Psychology, an MA in Ecology and PhD in Systems Ecology/Biology.

Contents

Preface

This is a book about water quality. Like all books concerning water quality, it is not comprehensive and does not cover all possible water quality topics. *Water Quality – Science, Assessments and Policy* is comprised of nine chapters examining scientific issues; national, regional and local assessment practices and results; and national policy issues. It is organized into three sections dealing with water quality parameters, water quality treatments, and water quality assessments. The two chapters on science issues examine the basic indicators of water quality and a unique sociological indicator of water quality. The three chapters covering water quality treatment evaluate natural water treatment for soil-transmitted helminths and ecological risk assessment using a sedimentary approach. Finally, the four chapters dealing with water quality assessment examine the US lakes, streams, estuaries and wetlands.

The section on water quality indicators includes two chapters. The first, "Water Quality Parameters," summarizes the water quality parameters in an ecological theme not only for humans but also for other living things. This chapter maintains that water can be classified into four types according to its quality. Those water quality types are discussed and an extensive review of the important common physical, chemicals and biological indicators is provided. The indicators are reviewed in terms of definition, sources, impacts, effects and measuring methods. The final chapter in this section, "Sense of Place and Water Quality: Applying Sense of Place Metrics to Better Understand Community Impacts of Changes in Water Quality," examines a sociological indicator of water quality – sense of place. This chapter focuses on understanding people's values for coastal and freshwater systems. This focus is critical for protecting water resources and informing management decisions. Sense of place is a social indicator that captures the relative value that different people hold for specific places and offers promise as a tool for measuring an important aspect of the social value of water quality. This chapter proposes a quantitative sense-of-place scale and additional qualitative questions that can be used in conjunction with biophysical water quality data and water quality perceptions data to better understand how people's values change with improvements or degradations in water quality.

The section on water quality treatments includes three chapters. The first chapter, "Natural Wastewater Treatment Systems for Prevention and Control of Soil-Transmitted Helminths," examines exposures to wastewater and their association to greater prevalence of soil-transmitted helminths. Despite preventive chemotherapy, wastewaters in many countries still contain high concentrations of soil-transmitted helminth eggs that put exposed populations at risk of infection. This chapter explains the role of natural wastewater treatment systems as sustainable sanitation facilities in removing STH from wastewater and therefore preventing disease transmission. The second chapter, "Association of Polyethylene Glycol Solubility with Emerging Membrane Technologies, Wastewater Treatment and Desalination," examines membrane technologies for the treatment of wastewater. In addition, this chapter evaluates the use of membrane technology for desalination. The final chapter in this section, "Water Quality Ecological Risk Assessment with Sedimentological Approach," looks at the creation of the potential ecological risk index (ERI) and its use as a diagnostic tool for water system assessment. The index is based on sedimentology and combined with environmental chemistry and ecotoxicology. The chapter introduces the index approach and uses it to assess the Liaohe River, China.

The section on water quality assessments includes four chapters. The 1972 Clean Water Act (CWA) established goals and regulations regarding water quality in US water resources, including coastal waters. The US Environmental Protection Agency (EPA) was charged with implementing the CWA's goals and with helping states and tribes meet their mandate to periodically monitor and assess water quality in their jurisdictions. In response, the EPA initiated the Environmental Monitoring and Assessment Program (EMAP) to research effective methods of assessing water quality in lakes, rivers and streams, and estuaries at state and national scales. Subsequently, the EMAP assessments evolved into the National Resource Surveys. All the chapters in this section relate to the US National Aquatic Resources Surveys and assess lakes, rivers and streams, coastal resources (estuaries and the Great Lakes), and wetlands. The lakes' assessments chapter, "Jewels across the Landscape: Monitoring and Assessing the Quality of Lakes and Reservoirs in the United States," describes efforts by a unique partnership between the United States and the EPA to monitor and assess lake ecosystems. The chapter on rivers and streams, "Rivers and Streams - Upgrading Monitoring of the Nation's Freshwater: Meeting the Spirit of the Clean Water Act," describes the partnership between EPA, the United States and tribes to remedy the information gap for rivers and streams. Filling this gap requires improved monitoring designs to reflect conditions across all waters as well as the expansion of indicators to move beyond water chemistry to include all three elements of the CWA: chemical, physical and biological integrity. The chapter on coastal resources, "Lessons Learned from 30 Years of Assessing U.S. Coastal Water," recounts the history of assessments in coastal waters, emphasizing the current approach while highlighting examples of lessons learned over the thirty-year development period leading to the National Coastal Condition Assessment. The final chapter is this section, "Wetland Assessment: Beyond the Traditional Water Quality Perspective," assesses the wetlands of the United States. The wetland survey introduces the concept of aquatic resource quality, the condition of an ecosystem based on the collective assessment of physical, chemical, and biological indicators, as the goal of monitoring and assessment of aquatic systems. The survey reports on wetland condition using a biotic indicator (vegetation multimetric index) and the relative extent and relative risk of stressors using ten physical, chemical and biological indicators. These surveys go beyond single water quality (chemistry) issues and include assessments targeting the goal of the CWA, namely, restoring, maintaining and protecting the chemical, physical and biological integrity of the nation's aquatic resources.

Water Quality – Science, Assessments and Policy provides basic understanding of water quality issues and practical examples of their solution.

Kevin Summers
United States Environmental Protection Agency,
Gulf Ecosystem Measurement and Modeling Division,
Florida, USA

Section 1

Water Quality Parameters

Chapter 1

Water Quality Parameters

Nayla Hassan Omer

Abstract

Since the industrial revolution in the late eighteenth century, the world has discovered new sources of pollution nearly every day. So, air and water can potentially become polluted everywhere. Little is known about changes in pollution rates. The increase in water-related diseases provides a real assessment of the degree of pollution in the environment. This chapter summarizes water quality parameters from an ecological perspective not only for humans but also for other living things. According to its quality, water can be classified into four types. Those four water quality types are discussed through an extensive review of their important common attributes including physical, chemical, and biological parameters. These water quality parameters are reviewed in terms of definition, sources, impacts, effects, and measuring methods.

Keywords: water quality, physical parameters, chemical parameters, biological parameters, radioactive substances, toxic substances, indicator organisms

1. Introduction

Water is the second most important need for life to exist after air. As a result, water quality has been described extensively in the scientific literature. The most popular definition of water quality is "it is the physical, chemical, and biological characteristics of water" [1, 2]. Water quality is a measure of the condition of water relative to the requirements of one or more biotic species and/or to any human need or purpose [3, 4].

2. Classification of water

Based on its source, water can be divided into ground water and surface water [5]. Both types of water can be exposed to contamination risks from agricultural, industrial, and domestic activities, which may include many types of pollutants such as heavy metals, pesticides, fertilizers, hazardous chemicals, and oils [6].

Water quality can be classified into four types—potable water, palatable water, contaminated (polluted) water, and infected water [7]. The most common scientific definitions of these types of water quality are as follows:

1. *Potable water:* It is safe to drink, pleasant to taste, and usable for domestic purposes [1, 7].

2. *Palatable water:* It is esthetically pleasing; it considers the presence of chemicals that do not cause a threat to human health [7].

3. *Contaminated (polluted) water:* It is that water containing unwanted physical, chemical, biological, or radiological substances, and it is unfit for drinking or domestic use [7].

4. *Infected water:* It is contaminated with pathogenic organism [7].

3. Parameters of water quality

There are three types of water quality parameters physical, chemical, and biological [8, 9]. They are summarized in **Table 1**.

3.1 Physical parameters of water quality

3.1.1 Turbidity

Turbidity is the cloudiness of water [10]. It is a measure of the ability of light to pass through water. It is caused by suspended material such as clay, silt, organic material, plankton, and other particulate materials in water [2].

No.	Types of water quality parameters		
	Physical parameters	**Chemical parameters**	**Biological parameters**
1	Turbidity	pH	Bacteria
2	Temperature	Acidity	Algae
3	Color	Alkalinity	Viruses
4	Taste and odor	Chloride	Protozoa
5	Solids	Chlorine residual	
6	Electrical conductivity (EC)	Sulfate	
7		Nitrogen	
8		Fluoride	
9		Iron and manganese	
10		Copper and zinc	
11		Hardness	
12		Dissolved oxygen	
13		Biochemical oxygen demand (BOD)	
14		Chemical oxygen demand (COD)	
15		Toxic inorganic substances	
16		Toxic organic substances	
17		Radioactive substances	

Table 1.
Parameters of water quality.

Turbidity in drinking water is esthetically unacceptable, which makes the water look unappetizing. The impact of turbidity can be summarized in the following points:

1. It can increase the cost of water treatment for various uses [11].

2. The particulates can provide hiding places for harmful microorganisms and thereby shield them from the disinfection process [12].

3. Suspended materials can clog or damage fish gills, decreasing its resistance to diseases, reducing its growth rates, affecting egg and larval maturing, and affecting the efficiency of fish catching method [13, 14].

4. Suspended particles provide adsorption media for heavy metals such as mercury, chromium, lead, cadmium, and many hazardous organic pollutants such as polychlorinated biphenyls (PCBs), polycyclic aromatic hydrocarbons (PAHs), and many pesticides [15].

5. The amount of available food is reduced [15] because higher turbidity raises water temperatures in light of the fact that suspended particles absorb more sun heat. Consequently, the concentration of the dissolved oxygen (DO) can be decreased since warm water carries less dissolved oxygen than cold water.

Turbidity is measured by an instrument called nephelometric turbidimeter, which expresses turbidity in terms of NTU or TU. A TU is equivalent to 1 mg/L of silica in suspension [10].

Turbidity more than 5 NTU can be visible to the average person while turbidity in muddy water, it exceeds 100 NTU [10]. Groundwater normally has very low turbidity because of the natural filtration that occurs as the water penetrates through the soil [9, 16].

3.1.2 Temperature

Palatability, viscosity, solubility, odors, and chemical reactions are influenced by temperature [10]. Thereby, the sedimentation and chlorination processes and biological oxygen demand (BOD) are temperature dependent [11]. It also affects the biosorption process of the dissolved heavy metals in water [17, 18]. Most people find water at temperatures of 10–15°C most palatable [10, 19].

3.1.3 Color

Materials decayed from organic matter, namely, vegetation and inorganic matter such as soil, stones, and rocks impart color to water, which is objectionable for esthetic reasons, not for health reasons [10, 20].

Color is measured by comparing the water sample with standard color solutions or colored glass disks [10]. One color unit is equivalent to the color produced by a 1 mg/L solution of platinum (potassium chloroplatinate (K_2PtCl_6)) [10].

The color of a water sample can be reported as follows:

- *Apparent color* is the entire water sample color and consists of both dissolved and suspended components color [10].

- *True color* is measured after filtering the water sample to remove all suspended material [19].

Color is graded on scale of 0 (clear) to 70 color units. Pure water is colorless, which is equivalent to 0 color units [10].

3.1.4 Taste and odor

Taste and odor in water can be caused by foreign matter such as organic materials, inorganic compounds, or dissolved gasses [19]. These materials may come from natural, domestic, or agricultural sources [21].

The numerical value of odor or taste is determined quantitatively by measuring a volume of sample A and diluting it with a volume of sample B of an odor-free distilled water so that the odor of the resulting mixture is just detectable at a total mixture volume of 200 ml [19, 22]. The unit of odor or taste is expressed in terms of a threshold number as follows:

$$\text{TON or TTN} = (A + B)/A \tag{1}$$

where TON is the threshold odor number and TTN is the threshold taste number.

3.1.5 Solids

Solids occur in water either in solution or in suspension [22]. These two types of solids can be identified by using a glass fiber filter that the water sample passes through [22]. By definition, the suspended solids are retained on the top of the filter and the dissolved solids pass through the filter with the water [10].

If the filtered portion of the water sample is placed in a small dish and then evaporated, the solids as a residue. This material is usually called total dissolved solids or TDS [10].

$$\text{Total solid (TS)} = \text{Total dissolved solid (TDS)} + \text{Total suspended solid (TSS)} \tag{2}$$

Water can be classified by the amount of TDS per liter as follows:

- freshwater: <1500 mg/L TDS;
- brackish water: 1500–5000 mg/L TDS;
- saline water: >5000 mg/L TDS.

The residue of TSS and TDS after heating to dryness for a defined period of time and at a specific temperature is defined as fixed solids. Volatile solids are those solids lost on ignition (heating to 550°C) [10].

These measures are helpful to the operators of the wastewater treatment plant because they roughly approximate the amount of organic matter existing in the total solids of wastewater, activated sludge, and industrial wastes [1, 22]. **Figure 1** describes the interrelationship of solids found in water [22]. They are calculated as follows [10]:

- Total solids:

$$\text{Total solids (mg/L)} = [(\text{TSA–TSB})] \times 1000/\text{sample (mL)} \tag{3}$$

where TSA = weight of dried residue + dish in milligrams and TSB = weight of dish in milligrams.

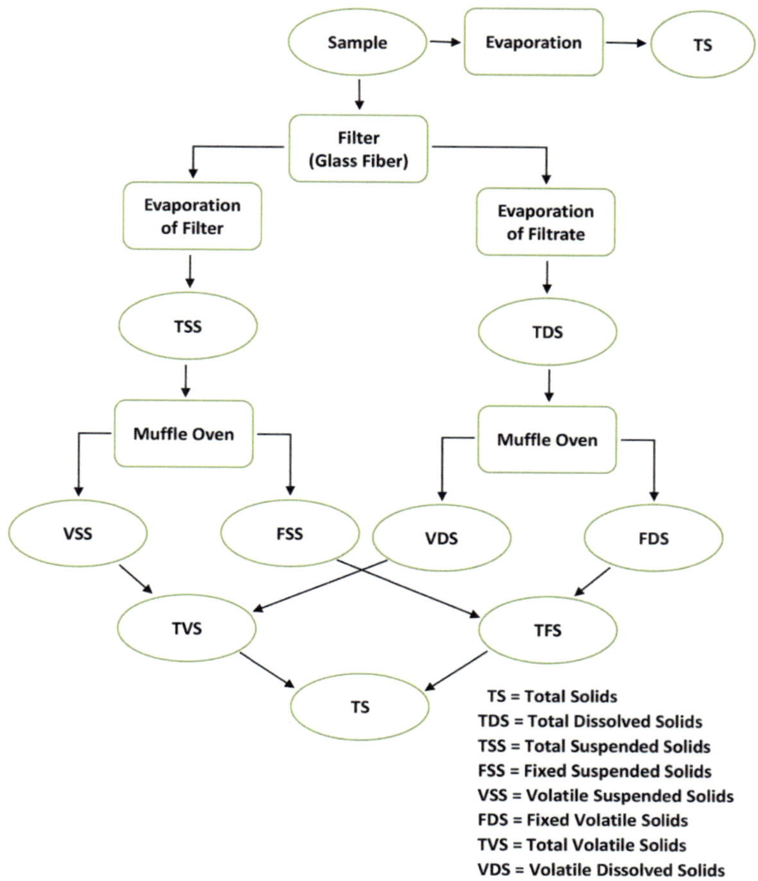

Figure 1.
Interrelationship of solids found in water [22].

- Total dissolved solids:

$$\text{Total dissolved solids (mg/L)} = [(\text{TDSA} - \text{TDSB})] \times 1000/\text{sample (mL)} \quad (4)$$

where TDSA = weight of dried residue + dish in milligrams and TDSB = weight of dish in milligrams.

- Total suspended solids:

$$\text{Total suspended solids (mg/L)} = [(\text{TSSA} - \text{TSSB})] \times 1000/\text{sample (mL)} \quad (5)$$

where TSSA = weight of dish and filter paper + dried residue and TSSB = weight of dish and filter paper in milligram.

- Fixed and volatile suspended solids:

$$\text{Volatile suspended solids (mg/L)} = [(\text{VSSA} - \text{VSSB})] \times 1000/\text{sample (mL)} \quad (6)$$

where VSSA = weight of residue + dish and filter before ignition, mg and VSSB = weight of residue + dish and filter after ignition, mg.

3.1.6 Electrical conductivity (EC)

The electrical conductivity (EC) of water is a measure of the ability of a solution to carry or conduct an electrical current [22]. Since the electrical current is carried by ions in solution, the conductivity increases as the concentration [10] of ions increases. Therefore, it is one of the main parameters used to determine the suitability of water for irrigation and firefighting.

Units of its measurement are as follows:

- U.S. units = micromhos/cm
- S.I. units = milliSiemens/m (mS/m) or dS/m (deciSiemens/m)

where (mS/m) = 10 umho/cm (1000 μS/cm = 1 dS/m).

Pure water is not a good conductor of electricity [2, 10]. Typical conductivity of water is as follows:

- Ultra-pure water: 5.5×10^{-6} S/m;
- Drinking water: 0.005–0.05 S/m;
- Seawater: 5 S/m.

The electrical conductivity can be used to estimate the TDS value of water as follows [10, 22]:

$$\text{TDS (mg/L)} \cong \text{EC (dS/m or umho/cm)} \times (0.55\text{–}0.7) \tag{7}$$

TDS can be used to estimate the ionic strength of water in the applications of groundwater recharging by treated wastewater [22]. The normal method of measurement is electrometric method [10].

3.2 Chemical parameters of water quality

3.2.1 pH

pH is one of the most important parameters of water quality. It is defined as the negative logarithm of the hydrogen ion concentration [9, 12]. It is a dimensionless number indicating the strength of an acidic or a basic solution [23]. Actually, pH of water is a measure of how acidic/basic water is [19, 20]. Acidic water contains extra hydrogen ions (H^+) and basic water contains extra hydroxyl (OH^-) ions [2].

As shown in **Figure 2**, pH ranges from 0 to 14, with 7 being neutral. pH of less than 7 indicates acidity, whereas a pH of greater than 7 indicates a base solution [2, 24]. Pure water is neutral, with a pH close to 7.0 at 25°C. Normal rainfall has a pH of approximately 5.6 (slightly acidic) owing to atmospheric carbon dioxide gas [10]. Safe ranges of pH for drinking water are from 6.5 to 8.5 for domestic use and living organisms need [24].

A change of 1 unit on a pH scale represents a 10-fold change in the pH [10], so that water with pH of 7 is 10 times more acidic than water with a pH of 8, and water with a pH of 5 is 100 times more acidic than water with a pH of 7. There are two methods available for the determination of pH: electrometric and colorimetric methods [10].

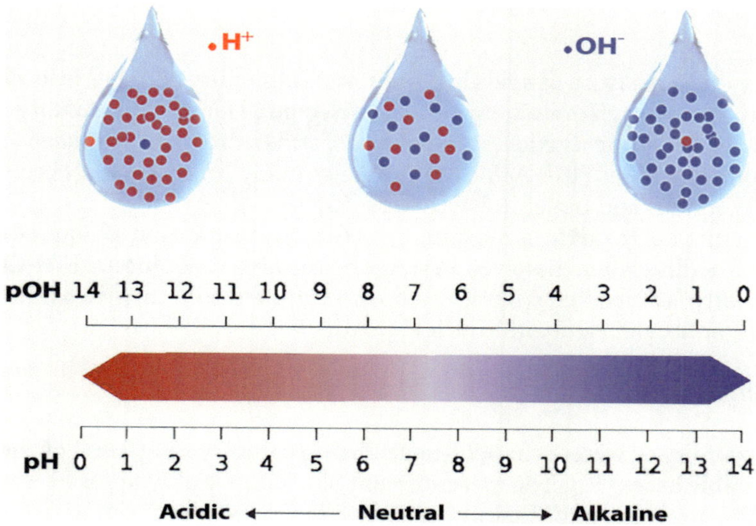

Figure 2.
pH of water.

Excessively high and low pHs can be detrimental for the use of water. A high pH makes the taste bitter and decreases the effectiveness of the chlorine disinfection, thereby causing the need for additional chlorine [21]. The amount of oxygen in water increases as pH rises. Low-pH water will corrode or dissolve metals and other substances [10].

Pollution can modify the pH of water, which can damage animals and plants that live in the water [10].

The effects of pH on animals and plants can be summarized as follows:

- Most aquatic animals and plants have adapted to life in water with a specific pH and may suffer from even a slight change [15].
- Even moderately acidic water (low pH) can decrease the number of hatched fish eggs, irritate fish and aquatic insect gills, and damage membranes [14].
- Water with very low or high pH is fatal. A pH below 4 or above 10 will kill most fish, and very few animals can endure water with a pH below 3 or above 11 [15].
- Amphibians are extremely endangered by low pH because their skin is very sensitive to contaminants [15]. Some scientists believe that the current decrease in amphibian population throughout the globe may be due to low pH levels induced by acid rain.

The effects of pH on other chemicals in water can be summarized as follows:

- Heavy metals such as cadmium, lead, and chromium dissolve more easily in highly acidic water (lower pH). This is important because many heavy metals become much more toxic when dissolved in water [21].
- A change in the pH can change the forms of some chemicals in the water. Therefore, it may affect aquatic plants and animals [21]. For instance, ammonia is relatively harmless to fish in neutral or acidic water. However, as the water becomes more alkaline (the pH increases), ammonia becomes progressively more poisonous to these same organisms.

3.2.2 Acidity

Acidity is the measure of acids in a solution. The acidity of water is its quantitative capacity to neutralize a strong base to a selected pH level [10]. Acidity in water is usually due to carbon dioxide, mineral acids, and hydrolyzed salts such as ferric and aluminum sulfates [10]. Acids can influence many processes such as corrosion, chemical reactions and biological activities [10].

Carbon dioxide from the atmosphere or from the respiration of aquatic organisms causes acidity when dissolved in water by forming carbonic acid (H_2CO_3). The level of acidity is determined by titration with standard sodium hydroxide (0.02 N) using phenolphthalein as an indicator [10, 20].

3.2.3 Alkalinity

The alkalinity of water is its acid-neutralizing capacity comprised of the total of all titratable bases [10]. The measurement of alkalinity of water is necessary to determine the amount of lime and soda needed for water softening (e.g., for corrosion control in conditioning the boiler feed water) [22]. Alkalinity of water is mainly caused by the presence of hydroxide ions (OH^-), bicarbonate ions (HCO^{3-}), and carbonate ions (CO_3^{2-}), or a mixture of two of these ions in water. As stated in the following equation, the possibility of OH^- and HCO_3^- ions together are not possible because they react together to produce CO_3^{2-} ions:

$$OH^- + HCO_3^- \rightarrow CO_3^{2-} + H_2O \tag{8}$$

Alkalinity is determined by titration with a standard acid solution (H_2SO_4 of 0.02 N) using selective indicators (methyl orange or phenolphthalein).

The high levels of either acidity or alkalinity in water may be an indication of industrial or chemical pollution. Alkalinity or acidity can also occur from natural sources such as volcanoes. The acidity and alkalinity in natural waters provide a buffering action that protects fish and other aquatic organisms from sudden changes in pH. For instance, if an acidic chemical has somehow contaminated a lake that had natural alkalinity, a neutralization reaction occurs between the acid and alkaline substances; the pH of the lake water remains unchanged. For the protection of aquatic life, the buffering capacity should be at least 20 mg/L as calcium carbonate.

3.2.4 Chloride

Chloride occurs naturally in groundwater, streams, and lakes, but the presence of relatively high chloride concentration in freshwater (about 250 mg/L or more) may indicate wastewater pollution [7]. Chlorides may enter surface water from several sources including chloride-containing rock, agricultural runoff, and wastewater.

Chloride ions Cl^- in drinking water do not cause any harmful effects on public health, but high concentrations can cause an unpleasant salty taste for most people. Chlorides are not usually harmful to people; however, the sodium part of table salt has been connected to kidney and heart diseases [25]. Small amounts of chlorides are essential for ordinary cell functions in animal and plant life.

Sodium chloride may impart a salty taste at 250 mg/L; however, magnesium or calcium chloride are generally not detected by taste until reaching levels of

1000 mg/L [10]. Standards for public drinking water require chloride levels that do not exceed 250 mg/L. There are many methods to measure the chloride concentration in water, but the normal one is the titration method by silver nitrate [10].

3.2.5 Chlorine residual

Chlorine (Cl_2) does not occur naturally in water but is added to water and wastewater for disinfection [10]. While chlorine itself is a toxic gas, in dilute aqueous solution, it is not harmful to human health. In drinking water, a residual of about 0.2 mg/L is optimal. The residual concentration which is maintained in the water distribution system ensures good sanitary quality of water [11].

Chlorine can react with organics in water forming toxic compounds called trihalomethanes or THMs, which are carcinogens such as chloroform $CHCl_3$ [11, 22]. Chlorine residual is normally measured by a color comparator test kit or spectrophotometer [10].

3.2.6 Sulfate

Sulfate ions (SO_4^{2-}) occur in natural water and in wastewater. The high concentration of sulfate in natural water is usually caused by leaching of natural deposits of sodium sulfate (Glauber's salt) or magnesium sulfate (Epson salt) [11, 26]. If high concentrations are consumed in drinking water, there may be objectionable tastes or unwanted laxative effects [26], but there is no significant danger to public health.

3.2.7 Nitrogen

There are four forms of nitrogen in water and wastewater: organic nitrogen, ammonia nitrogen, nitrite nitrogen, and nitrate nitrogen [10]. If water is contaminated with sewage, most of the nitrogen is in the forms of organic and ammonia, which are transformed by microbes to form nitrites and nitrates [22]. Nitrogen in the nitrate form is a basic nutrient to the growth of plants and can be a growth-limiting nutrient factor [10].

A high concentration of nitrate in surface water can stimulate the rapid growth of the algae which degrades the water quality [22]. Nitrates can enter the groundwater from chemical fertilizers used in the agricultural areas [22]. Excessive nitrate concentration (more than 10 mg/L) in drinking water causes an immediate and severe health threat to infants [19]. The nitrate ions react with blood hemoglobin, thereby reducing the blood's ability to hold oxygen which leads to a disease called blue baby or methemoglobinemia [10, 19].

3.2.8 Fluoride

A moderate amount of fluoride ions (F^-) in drinking water contributes to good dental health [10, 19]. About 1.0 mg/L is effective in preventing tooth decay, particularly in children [10].

Excessive amounts of fluoride cause discolored teeth, a condition known as dental fluorosis [11, 19, 26]. The maximum allowable levels of fluoride in public water supplies depend on local climate [26]. In the warmer regions of the country, the maximum allowable concentration of fluoride for potable water is 1.4 mg/L; in colder climates, up to 2.4 mg/L is allowed.

There are four methods to determine ion fluoride in water; the selection of the used method depends on the type of water sample [10].

3.2.9 Iron and manganese

Although iron (Fe) and manganese (Mn) do not cause health problems, they impart a noticeable bitter taste to drinking water even at very low concentration [10, 11].

These metals usually occur in groundwater in solution as ferrous (Fe^{2+}) and manganous (Mn^{2+}) ions. When these ions are exposed to air, they form the insoluble ferric (Fe^{3+}) and manganic (Mn^{3+}) forms making the water turbid and unacceptable to most people [10].

These ions can also cause black or brown stains on laundry and plumbing fixtures [7]. They are measured by many instrumental methods such as atomic absorption spectrometry, flame atomic absorption spectrometry, cold vapor atomic absorption spectrometry, electrothermal atomic absorption spectrometry, and inductively coupled plasma (ICP) [10].

3.2.10 Copper and zinc

Copper (Cu) and zinc (Zn) are nontoxic if found in small concentrations [10]. Actually, they are both essential and beneficial for human health and growth of plants and animals [25]. They can cause undesirable tastes in drinking water. At high concentrations, zinc imparts a milky appearance to the water [10]. They are measured by the same methods used for iron and manganese measurements [10].

3.2.11 Hardness

Hardness is a term used to express the properties of highly mineralized waters [10]. The dissolved minerals in water cause problems such as scale deposits in hot water pipes and difficulty in producing lather with soap [11].

Calcium (Ca^{2+}) and magnesium (Mg^{2+}) ions cause the greatest portion of hardness in naturally occurring waters [9]. They enter water mainly from contact with soil and rock, particularly limestone deposits [10, 27].

These ions are present as bicarbonates, sulfates, and sometimes as chlorides and nitrates [10, 26]. Generally, groundwater is harder than surface water. There are two types of hardness:

- *Temporary hardness* which is due to carbonates and bicarbonates can be removed by boiling, and

- *Permanent hardness* which is remaining after boiling is caused mainly by sulfates and chlorides [10, 21, 22]

Water with more than 300 mg/L of hardness is generally considered to be hard, and more than 150 mg/L of hardness is noticed by most people, and water with less than 75 mg/L is considered to be soft.

From health viewpoint, hardness up to 500 mg/L is safe, but more than that may cause a laxative effect [10]. Hardness is normally determined by titration with ethylene diamine tetra acidic acid or (EDTA) and Eriochrome Black and Blue indicators. It is usually expressed in terms of mg/L of $CaCO_3$ [10, 19].

$$\text{Total hardness mg/L as } CaCO_3 = \text{calcium hardness mg/L as } CaCO_3 + \text{magnesium hardness mg/L as } CaCO_3 \quad (9)$$

An accepted water classification according to its hardness is as in **Table 2** [19].

3.2.12 Dissolved oxygen

Dissolved oxygen (DO) is considered to be one of the most important parameters of water quality in streams, rivers, and lakes. It is a key test of water pollution [10]. The higher the concentration of dissolved oxygen, the better the water quality.

Oxygen is slightly soluble in water and very sensitive to temperature. For example, the saturation concentration at 20°C is about 9 mg/L and at 0°C is 14.6 mg/L [22].

The actual amount of dissolved oxygen varies depending on pressure, temperature, and salinity of the water. Dissolved oxygen has no direct effect on public health, but drinking water with very little or no oxygen tastes unpalatable to some people.

There are three main methods used for measuring dissolved oxygen concentrations: the colorimetric method—quick and inexpensive, the Winkler titration method—traditional method, and the electrometric method [10].

3.2.13 Biochemical oxygen demand (BOD)

Bacteria and other microorganisms use organic substances for food. As they metabolize organic material, they consume oxygen [10, 22]. The organics are broken down into simpler compounds, such as CO_2 and H_2O, and the microbes use the energy released for growth and reproduction [22].

When this process occurs in water, the oxygen consumed is the DO in the water. If oxygen is not continuously replaced by natural or artificial means in the water, the DO concentration will reduce as the microbes decompose the organic materials. This need for oxygen is called the biochemical oxygen demand (BOD). The more organic material there is in the water, the higher the BOD used by the microbes will be. BOD is used as a measure of the power of sewage; strong sewage has a high BOD and weak sewage has low BOD [22].

The complete decomposition of organic material by microorganisms takes time, usually 20 d or more under ordinary circumstances [22]. The quantity of oxygen used in a specified volume of water to fully decompose or stabilize all biodegradable organic substances is called the ultimate BOD or BOD_L.

BOD is a function of time. At time = 0, no oxygen will have been consumed and the BOD = 0. As each day goes by, oxygen is used by the microbes and the BOD increases. Ultimately, the BOD_L is reached and the organic materials are completely decomposed.

Water classification	Total hardness concentration as mg/L as $CaCO_3$
Soft water	<50 mg/L as $CaCO_3$
Moderately hard	50–150 mg/L as $CaCO_3$
Hard water	150–300 mg/L as $CaCO_3$
Very hard	>300 mg/L as $CaCO_3$

Table 2.
Classification of water according to its hardness.

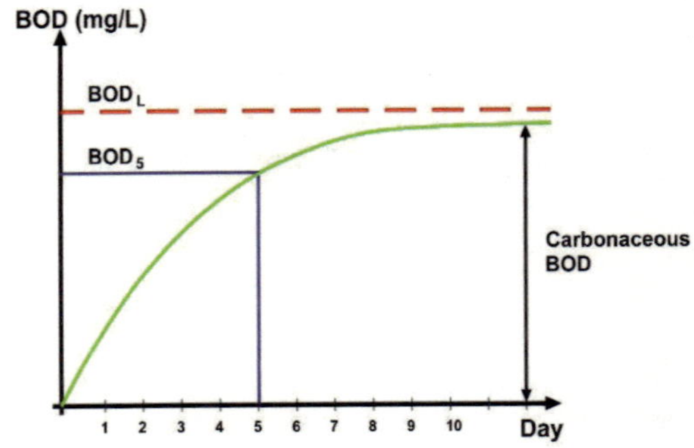

Figure 3.
BOD curve [22].

A graph of the BOD versus time is illustrated as in **Figure 3**. This is called the BOD curve, which can be expressed mathematically by the following equation:

$$BOD_t = BOD_L \times \left(1 - 10^{-kt}\right) \tag{10}$$

where BOD_t = BOD at any time t, mg/L; BOD_L = ultimate BOD, mg/L; k = a constant representing the rate of the BOD reaction; t = time, d.

The value of the constant rate k depends on the temperature, the type of organic materials, and the type of microbes exerting the BOD [22].

3.2.14 Chemical oxygen demand (COD)

The chemical oxygen demand (COD) is a parameter that measures all organics: the biodegradable and the non-biodegradable substances [22]. It is a chemical test using strong oxidizing chemicals (potassium dichromate), sulfuric acid, and heat, and the result can be available in just 2 h [10]. COD values are always higher than BOD values for the same sample [22].

3.2.15 Toxic inorganic substances

A wide variety of inorganic toxic substances may be found in water in very small or trace amounts. Even in trace amounts, they can be a danger to public health [11]. Some toxic substances occur from natural sources but many others occur due to industrial activities and/or improper management of hazardous waste [22]. They can be divided into two groups:

- *Metallic compounds:* This group includes some heavy metals that are toxic, namely, cadmium (Cd), chromium (Cr), lead (Pb), mercury (Hg), silver (Ag), arsenic (As), barium (Ba), thallium (Tl), and selenium (Se) [22, 28]. They have a wide range of dangerous effects that differ from one metal to another. They may be acute fatal poisons such as (As) and (Cr^{6+}) or may produce chronic diseases such as (Cd, Hg, Pb, and Tl) [21, 29–32]. The heavy metals concentration can be determined by atomic absorption photometers, spectrophotometer, or inductively coupled plasma (ICP) for very low concentration [10].

- *Nonmetallic compounds:* This group includes nitrates (NO_3^-) and cyanides (CN^-), nitrate has been discussed with the nitrogen in the previous section. Regarding cyanide, as Mackenzie stated [11] it causes oxygen deprivation by binding the hemoglobin sites and prevents the red blood cell from carrying the oxygen [11]. This causes a blue skin color syndrome, which is called cyanosis [33]. It also causes chronic effects on the central nervous system and thyroid [33]. Cyanide is normally measured by colorimetric, titrimetric, or electrometric methods [10].

3.2.16 Toxic organic substances

There are more than 100 compounds in water that have been listed in the literature as toxic organic compounds [11, 22]. They will not be found naturally in water; they are usually man-made pollutants. These compounds include insecticides, pesticides, solvents, detergents, and disinfectants [11, 21, 22]. They are measured by highly sophisticated instrumental methods, namely, gas chromatographic (GC), high-performance liquid chromatographic (HPLC), and mass spectrophotometric [10].

3.2.17 Radioactive substances

Potential sources of radioactive substances in water include wastes from nuclear power plants, industries, or medical research using radioactive chemicals and mining of uranium ores or other radioactive materials [11, 21]. When radioactive substances decay, they release beta, alpha, and gamma radiation [34]. Exposure of humans and other living things to radiation can cause genetic and somatic damage to the living tissues [34, 35].

Radon gas is of a great health concern because it occurs naturally in groundwater and is a highly volatile gas, which can be inhaled during the showering process [35]. For drinking water, there are established standards commonly used for alpha particles, beta particles, photons emitters, radium-226 and -228, and uranium [34, 35].

The unit of radioactivity used in water quality applications is the picocurie per liter (pCi/L); 1 pCi is equivalent to about two atoms disintegrating per minute. There are many sophisticated instrumental methods to measure it [35].

3.3 Biological parameters of water quality

One of the most helpful indicators of water quality may be the presence or lack of living organisms [10, 15]. Biologists can survey fish and insect life of natural waters and assess the water quality on the basis of a computed species diversity index (SDI) [15, 19, 36, 37]; hence, a water body with a large number of well-balanced species is regarded as a healthy system [17]. Some organisms can be used as an indication for the existence of pollutants based on their known tolerance for a specified pollutant [17].

Microorganisms exist everywhere in nature [38]. Human bodies maintain a normal population of microbes in the intestinal tract; a big portion of which is made up of coliform bacteria [38]. Although there are millions of microbes per milliliter in wastewater, most of them are harmless [37]. It is only harmful when wastewater contains wastes from people infected with diseases that the presence of harmful microorganisms in wastewater is likely to occur [38].

3.3.1 Bacteria

Bacteria are considered to be single-celled plants because of their cell structure and the way they ingest food [10, 37]. Bacteria occur in three basic cell shapes: rod-shaped or bacillus, sphere-shaped or coccus, and spiral-shaped or spirellus [19]. In less than 30 min, a single bacterial cell can mature and divide into two new cells [39].

Under favorable conditions of food supply, temperature, and pH, bacteria can reproduce so rapidly that a bacterial culture may contain 20 million cells per milliliter after just 1 day [22, 37]. This rapid growth of visible colonies of bacteria on a suitable nutrient medium makes it possible to detect and count the number of bacteria in water [39].

There are several distinctions among the various species of bacteria. One distinction depends on how they metabolize their food [38]. Bacteria that require oxygen for their metabolism are called aerobic bacteria, while those live only in an oxygen-free environment are called anaerobic bacteria. Some species called facultative bacteria can live in either the absence or the presence of oxygen [37–39].

At low temperatures, bacteria grow and reproduce slowly. As the temperature increases, the rate of growth and reproduction doubles in every additional 10°C (up to the optimum temperature for the species) [38]. The majority of the species of bacteria having an optimal temperature of about 35°C [39].

A lot of dangerous waterborne diseases are caused by bacteria, namely, typhoid and paratyphoid fever, leptospirosis, tularemia, shigellosis, and cholera [19]. Sometimes, the absence of good sanitary practices results in gastroenteritis outbreaks of one or more of those diseases [19].

3.3.2 Algae

Algae are microscopic plants, which contain photosynthetic pigments, such as chlorophyll [37, 39]. They are autotrophic organisms and support themselves by converting inorganic materials into organic matter by using energy from the sun, during this process they take in carbon dioxide and give off oxygen [38, 39]. They are also important for wastewater treatment in stabilization ponds [22]. Algae are primarily nuisance organisms in the water supply because of the taste and odor problems they create [2, 16]. Certain species of algae cause serious environmental and public health problems; for example, blue-green algae can kill cattle and other domestic animals if the animals drink water containing those species [37, 39].

3.3.3 Viruses

Viruses are the smallest biological structures known to contain all genetic information necessary for their own reproduction [19]. They can only be seen by a powerful electronic microscope [39]. Viruses are parasites that need a host to live [39]. They can pass through filters that do not permit the passage of bacteria [37]. Waterborne viral pathogens are known to cause infectious hepatitis and poliomyelitis [19, 25, 37]. Most of the waterborne viruses can be deactivated by the disinfection process conducted in the water treatment plant [19].

3.3.4 Protozoa

Protozoa are single-celled microscopic animal [19], consume solid organic particles, bacteria, and algae for food, and they are in turn ingested as food by higher level multicellular animals [37]. Aquatic protozoa are floating freely in water and sometimes called zooplankton [37]. They form cysts that are difficult to inactivate by disinfection [19].

3.3.5 Indicator organisms

A very important biological indicator of water and pollution is the group of bacteria called coliforms [20]. Pathogenic coliforms always exist in the intestinal system of humans, and millions are excreted with body wastes [37]. Consequently, water that has been recently contaminated with sewage will always contain coliforms [19].

A particular species of coliforms found in domestic sewage is *Escherichia coli* or *E. coli* [22]. Even if the water is only slightly polluted, they are very likely to be found. There are roughly 3 million of *E. coli* bacteria in 100 mL volume of untreated sewage [10]. Coliform bacteria are aggressive organisms and survive in the water longer than most pathogens. There are normally two methods to test the coliform bacteria—the membrane filter method and multiple-tube fermentation method [10, 37]. Since the test of coliform bacteria is very important for public health, the first method will be described in details in the coming section.

3.3.5.1 Testing for coliforms: membrane filter method

A measured volume of sample is filtered through a special membrane filter by applying a partial vacuum [10, 39].

The filter, a flat paper-like disk, has uniform microscopic pores small enough to retain the bacteria on its surface while allowing the water to pass through. The filter paper is then placed in a sterile container called a petri dish, which contains a special culture medium that the bacteria use as a food source [39].

Then, the petri dish is usually placed in an incubator, which keeps the temperature at 35°C, for 24 h. After incubation, colonies of coliform bacteria each containing millions of organisms will be visible [10]. The coliform concentration is obtained by counting the number of colonies on the filter; each colony counted represents only one coliform in the original sample [10, 39].

Coliform concentrations are expressed in terms of the number of organisms per 100 mL of water as follows:

$$\text{coliforms per 100 mL} = \text{number of colonies} \times 100/\text{mL of sample} \tag{11}$$

4. Water quality requirements

Water quality requirements differ depending on the proposed used of water [19]. As reported by Tchobanoglous et al. [19], "water unsuitable for one use may be quite satisfactory for another and water may be considered acceptable for a particular use if water of better quality is not available."

Water quality requirements should be agreed with the water quality standards, which are put down by the governmental agency and represent the legislation requirements. In general, there are three types of standards: in-stream, potable water, and wastewater effluent [19], each type has its own criteria by using the same methods of measurement. The World Health Organization (WHO) has established minimum standards for drinking water that all countries are recommended to meet [25].

5. Conclusion

The physical, chemical, and biological parameters of water quality are reviewed in terms of definition, sources, impacts, effects, and measuring methods. The classification of water according to its quality is also covered with a specific definition for each type.

Author details

Nayla Hassan Omer
Department of Environmental Engineering, College of Water and Environmental Engineering, Sudan University for Science and Technology, Khartoum, Sudan

*Address all correspondence to: naylahomer@yahoo.com

References

[1] Spellman FR. Handbook of Water and Wastewater Treatment Plant Operations. 3rd ed. Boca Raton: CRC Press; 2013

[2] Alley ER. Water Quality Control Handbook. Vol. 2. New York: McGraw-Hill; 2007

[3] Shah C. Which Physical, Chemical and Biological Parameters of Water Determine Its Quality?; 2017

[4] Tchobanoglous G, Schroeder E. Water Quality: Characteristics, Modeling, Modification. 1985

[5] Gray N. Water Technology. 3rd ed. London: CRC Press; 2017

[6] Davis ML, Masten SJ. Principles of Environmental Engineering and Science. New York: McGraw-Hill; 2004

[7] Chatterjee A. Water Supply Waste Disposal and Environmental Pollution Engineering (Including Odour, Noise and Air Pollution and its Control). 7th ed. Delhi: Khanna Publishers; 2001

[8] Gray NF. Drinking Water Quality: Problems and Solutions. 2nd ed. Cambridge: Cambridge University Press; 2008

[9] Spellman FR. The Drinking Water Handbook. 3rd ed. Boca Raton: CRC Press; 2017

[10] APHA. Standard Methods for the Examination of Water and Wastewater. 21st ed. Washington, DC: American Public Health Association; 2005

[11] Davis ML. Water and Wastewater Engineering—Design Principles and Practice. New York: McGraw-Hill; 2010

[12] Edzwald JK. Water Quality and Treatment a Handbook on Drinking Water. New York: McGraw-Hill; 2010

[13] Tarras-Wahlberg H, Harper D, Tarras-Wahlberg N. A first limnological description of Lake Kichiritith, Kenya: A possible reference site for the freshwater lakes of the Gregory Rift valley. South African Journal of Science. 2003;**99**:494-496

[14] Kiprono SW. Fish Parasites and Fisheries Productivity in Relation to Exreme Flooding of Lake Baringo, Kenya [PhD]. Nairobi: Kenyatta University; 2017

[15] Cole S, Codling I, Parr W, Zabel T, Nature E, Heritage SN. Guidelines for Managing Water Quality Impacts within UK European Marine Sites; 1999

[16] Viessman W, Hammer MJ. Water Supply and Pollution Control. 7th ed. Upper Saddle River: New Jersey Pearson Prentice Hall; 2004

[17] Abbas SH, Ismail IM, Mostafa TM, Sulaymon AH. Biosorption of heavy metals: A review. Journal of Chemical Science and Technology. 2014;**3**:74-102

[18] White C, Sayer J, Gadd G. Microbial solubilization and immobilization of toxic metals: Key biogeochemical processes for treatment of contamination. FEMS Microbiology Reviews. 1997;**20**:503-516

[19] Tchobanoglous G, Peavy HS, Rowe DR. Environmental Engineering. New York: McGraw-Hill Interamericana; 1985

[20] Tomar M. Quality Assessment of Water and Wastewater. Boca Raton: CRC Press; 1999

[21] DeZuane J. Handbook of Drinking Water Quality. 2nd ed. New York: John Wiley & Sons; 1997

[22] Tchobanoglous G, Burton FL, Stensel HD. Metcalf & Eddy Wastewater Engineering: Treatment and Reuse.

4th ed. New Delhi: Tata McGraw-Hill Limited; 2003

[23] Hammer MJ. Water and Wastewater Technology. 7th ed. Upper Saddle River: Pearson education; 2011

[24] World Health Organization Guidelines for drinking-water quality. 4th ed. Geneva: WHO; 2011

[25] World Health Organization. Guidelines for Drinking-Water Q, Vol. 2, Health criteria and other supporting information. 1996

[26] Davis ML, David A. Introduction to Environmental Engineering. 4th ed. New York: McGraw-Hill Companies; 2008

[27] McGhee TJ, Steel EW. Water Supply and Sewerage. New York: McGraw-Hill; 1991

[28] Järup L. Hazards of heavy metal contamination. British Medical Bulletin. 2003;**68**:167-182

[29] Campanella B, Onor M, D'Ulivo A, Giannecchini R, D'Orazio M, Petrini R, et al. Human exposure to thallium through tap water: A study from Valdicastello Carducci and Pietrasanta (northern Tuscany, Italy). Science of the Total Environment. 2016;**548**:33-42

[30] Das AK, Dutta M, Cervera ML, de la Guardia M. Determination of thallium in water samples. Microchemical Journal. 2007;**86**:2-8

[31] Lasheen MR, Shehata SA, Ali GH. Effect of cadmium, copper and chromium (VI) on the growth of Nile water algae. Water, Air, & Soil Pollution. 1990;**50**:19-30

[32] World Health Organization. Chromium in Drinking-Water. Geneva: World Health Organization (WHO); 2003

[33] Dojlido J, Best GA. Chemistry of Water and Water Pollution. Chichester: Ellis Horwood Limited; 1993

[34] Skeppström K, Olofsson B. Uranium and radon in groundwater. European Water. 2007;**17**:51-62

[35] Cothern CR. Radon, Radium, and Uranium in Drinking Water. Boca Raton: CRC Press; 2014

[36] Alabaster JS, Lloyd RS. Water Quality Criteria for Freshwater Fish. 2nd ed. Cambridge: Butterworths; 1984

[37] Nathanson JA. Basic Environmental Technology: Water Supply. New Delhi: Printice-Hall of India; 2004

[38] Wiesmann U, Choi IS, Dombrowski E-M. Fundamentals of Biological Wastewater Treatment. Darmstadt: John Wiley & Sons; 2007

[39] Mara D, Horan NJ. Handbook of Water and Wastewater Microbiology. London: Elsevier; 2003

Chapter 2

Sense of Place and Water Quality: Applying Sense of Place Metrics to Better Understand Community Impacts of Changes in Water Quality

Kate K. Mulvaney, Nathaniel H. Merrill and Marisa J. Mazzotta

Abstract

Understanding people's values for coastal and freshwater areas is critical for identifying concerns and motivating people to protect water resources and for informing management decisions. Sense of place is a social indicator that captures the relative value that different people hold for specific places. Its use in water quality assessments remains extremely limited but based on lessons from other environmental fields, sense of place offers promise as a tool for measuring an important aspect of the social value of water quality. In this chapter, we propose a quantitative sense-of-place scale and additional qualitative questions which can be used in conjunction with biophysical water quality data and water quality perceptions data to better understand how people's values change with improvements or degradations in water quality.

Keywords: sense of place, water quality, social science, cultural ecosystem services

1. Introduction

Coastal and freshwater areas are important (or not) to people for a number of reasons ranging from the provision of the right resources for recreational activities to ease of accessibility for a family's use. Decisions about actions and policies that affect water quality can be better informed by understanding what makes people value various locations and how improvements or degradations in water quality can affect that value. While biophysical data are being increasingly collected, analyzed, and applied to critical environmental decisions, complementary social data remain relatively scarce. This presents a significant problem, as water quality impairments are inherently social problems in specific locations and effective solutions require public support and community willingness to make decisions and changes. Even the most readily available social data related to water quality—water quality perceptions and travel cost studies—are limited in scope and quantity and often do not consider the extent of environmental attributes of those places [1].

Sense of place has significant potential as an indicator of the social value of different locations and their environmental attributes. In this chapter, we focus on using sense of place to capture values related to water quality and connecting this sense of place value with biophysical water quality metrics and other social indicators. To date, sense of place assessments have focused on water quality have been conducted in only a few places and most do not link to specific biophysical metrics. We highlight the utility of sense of place as an indicator of the relative importance of different sites and its potential for assessing water quality in conjunction with other social and biophysical data. First, we review the literature on sense of place and its historical application and findings. We then describe the few existing applications of sense of place in the context of environmental attributes, including water quality. This chapter ends with a call for researchers to use sense of place as a cultural ecosystem service indicator and we present a proposed sense of place scale for use in water quality social assessments.

2. Sense of place

Sense of place analyses are mechanisms for articulating the social value of a geographical area. Specifically, "sense of place" is a social theory that connects an individual's meaning and attachment for a specific geographical place with the attributes of that place such as amenities, site characteristics, and environmental quality [2]. Sense of place can be a useful tool for quantifying and characterizing the social value of water quality as coastal and freshwater places are more than just their environmental attributes. These places provide important meaning and value to the people who inhabit and visit them.

In the application of sense of place, "place" is generally a specific geography that is defined based on political or natural boundaries or other special features [3]. Place is, importantly, identified in this context as not just a stage for social interactions but as a critical component of those interactions [4]. For example, one coastal place may be a neighborhood access point to a small estuary that is primarily used for launching kayaks or exploring tidepools. Another coastal place might be a larger beach visited by residents from multiple states that is operated by a state as a park and offers full amenities such parking, restrooms, and lifeguards.

Sense of place provides a useful indicator of social value, as the components of sense of place have been connected to increased community involvement as well as environmental protection responses, concern, intentions, and behavior (e.g., [5–11]). For example, Lukacs and Ardoin [12] connected sense of place with participation and motivation for engagement with local watershed management groups in Appalachia. Similarly, sense of place attitudes have been connected with behaviors such as opposition to new renewable energy development and natural protected areas, which were seen as threats to the autonomy and opportunities of the local residents [1, 13]. As the concept of sense of place inherently recognizes humans as a component of the ecosystem, it helps bridge the gap between scientific research and environmental decision making by elucidating some of the social value of environmental protection [14]. For example, in a Nebraska river watershed effort, Davenport and Anderson [15] developed a place-meanings framework that can be used by managers to better understand the complexities behind contentious issues.

The bulk of sense of place research has focused on individual-level attitudes toward a particular geographical area, but there is also a body of work connecting broader sociocultural values and perspectives to more general geographical constructs. Put more simply, an individual's interaction with a specific place does not exist in a vacuum separate from the broader geography or society. Larger socio-cultural constraints and interactions influence individual's or communities' emotional relationships to a place [16–18]. For example, Campbell [19] studied the shared sense of place values in Ontario's eastern Georgian Bay and found sub-communities (i.e., artists/writers and residents) developed a specific language and identity that was different from that of people in other areas within the region, revealing a unique reference to their home communities. Poe et al. [17] found that sense of place for residents of Puget Sound was multidimensional with the availability of access to the Sound, knowledge about use, access and conditions, and perceived ecological integrity influencing place attachment. The environmental characteristics of a location also matter, as sense of place is not just a social construct [20], and some research has supported the idea that people's sense of place can be for both a specific geography as well as for general places that share similar characteristics, including environmental attributes [21]. One example of this research [22] found some differences between site specific attachment to a wilderness location and more general attachment to wilderness areas.

3. The components of sense of place

Sense of place can be considered broadly as an overall measure and can be investigated through several subcomponents—most commonly place dependence, place identity, and place attachment (see **Figure 1**). Relationships with a site because of its functional provision of particular resources that support activities (e.g., waves for surfing, clarity for diving, bacteria-free access for swimming, or scenic vistas for viewing) are described as *place dependence* [22]. A person with high place dependence would ascribe high importance on the availability of a specific condition at a site relative to other sites [23]. Place dependence has been investigated in several

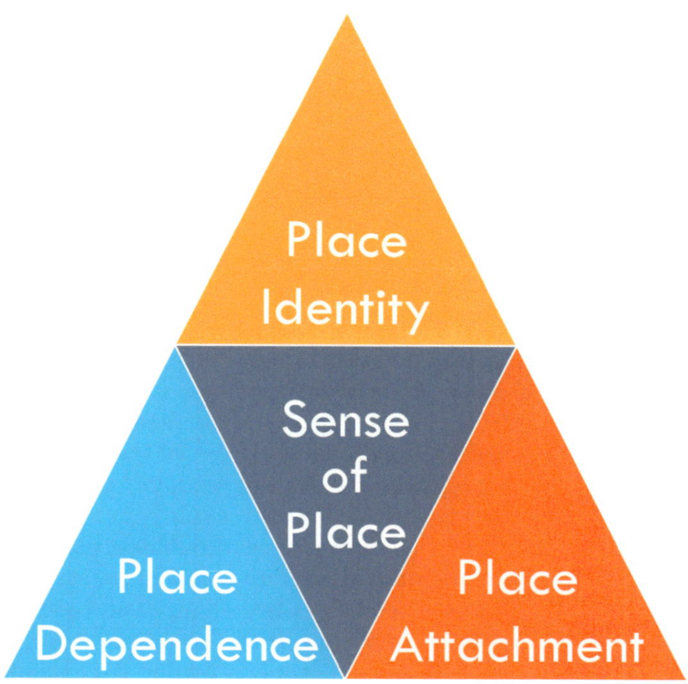

Figure 1.
Sense of place. Sense of place is often discussed in terms of three subcomponents: place dependence, place identity, and place attachment.

studies of various types of recreation areas and was found to increase with perceived familiarity with a park [23], and with length of residence and education [24].

Several studies have investigated hypotheses that sense of place (e.g., often place dependence, specifically) is different for various types of users but have not always found consistent differences [2]. For example, Bricker and Kerstetter [25] found that, for whitewater recreationists, how particularly specialized recreational users were described in terms of equipment, skills, and activity frequency did not affect place dependence, although it did influence place identity. When applying sense of place metrics to water quality assessments, degree of specialization is important because the acceptable type of recreation may vary considerably for different water quality conditions. For example, someone may still be willing to go for a walk on a beach that is closed to swimming because of bacterial contamination but would not be willing to go diving or swimming. Specialization also matters when connecting the social value of a place to economic values, as people tend to have different values depending on the type of recreational use, and users' individual attributes (e.g., more or less avid, residents or visitors) affect their values for attributes of a place.

Place identity can be described as the emotional counterpart to place dependence. Instead of measuring the dependence on a place for its resources to support an activity or livelihood, place identity captures the dependence on the place for constructing one's self-identity [2, 16]. The specific place builds symbolic importance for an individual's emotions and self-identification [26]. For example, people with high place identity for water areas might express sentiments like "I'm an ocean person," or "Water is a part of who I am." Higher place identity has been identified in those who are more familiar with a recreational site [23]; in those who have a higher degree of recreational specialization [25]; or in those who live in rural areas, have a longer time of residence, and own their homes [24].

The most studied of the main sub-components of sense of place is *place attachment*. In part this is because, out of the components, it has the broadest breadth of meanings [2, 27]. In some fields and research efforts, place attachment and sense of place are virtually synonymous [23]. In other applications, particularly in recreation-based work, place attachment is a subcomponent of sense of place that captures the emotional bond with a place or how important a place is to someone beyond the resource or identity dependences. Place attachment is rooted in Tuan's [28] seminal work on topophilia which focuses on the connection of an individual to a place.

Past studies have shown that local environmental perceptions, as well as the number of local social relationships, increase place attachment [8]. Someone with high attachment may have a lot of fond memories of visiting a place or consider themselves bonded with a location. There is a range of findings associated with place attachment to natural areas. Some researchers have found that place attachment to lands or sites is higher for locals (e.g., [1]). Others (e.g., [29]) have found that place attachment is higher for more repetitive users of the area.

A number of researchers (e.g., [5, 15, 20, 30]) have argued the importance of in sense of place is not just the strength of these three sub-components, but also the overall meanings an individual ascribes to a location. These place meanings are often captured through complementary quantitative/qualitative studies or standalone qualitative investigations into people's values, significance, and descriptions of the place [30]. A place may hold diverse meanings for different individuals, such as a recreationist at a site versus an adjacent property owner. For example, in the Midwest, Mullendore et al. [31] argue that farmers' sense of place values have not been captured in most sense of place studies. They explain that most sense of place studies have targeted recreational use in parks, wilderness, or other natural areas rather than exploring place values for working landscapes, which have different meanings associated with them. Place meanings can also have management implications. Jacobs and Buijs [32] found that attitudes toward water-level and restoration management interventions depended on the stakeholders' place meanings for the area.

While these three components are often talked about in terms of contributing to overall sense of place attitudes, there are not always clear-cut boundaries among them, and a number of researchers have identified varying relationships and interactions among them. For example, Kyle et al. [33] found conflicting effects between place identity and dependence for social and environmental conditions on the Appalachian Trail despite a moderate positive correlation of place identity and dependence. They found that respondents with higher place identity were more likely to see social and environmental conditions as problematic while the opposite was true for those with high place dependence. For Indiana farmers, Mullendore et al. [31] looked at the relationship between sense of place and willingness to adopt specific conservation behaviors in a working landscape (related to nutrient pollution in waters). They found the magnitude of the overall sense of place scale did not affect conservation adoption, but place attachment and place identity individually did.

4. Policy application of sense of place

Application of sense of place to policy questions is relatively new and there have been mixed findings about its utility for informing environmental management [2]. This may have resulted from sense of place studies not being conducted using consistent metrics or methods and the findings have also not always been consistent across places and studies. Researchers are increasingly applying similar quantitative scale questions, most particularly the scales developed by Jorgensen and Stedman [26] and Williams and Vaske [23]. Qualitative investigations of sense of place follow consistent themes investigating place attachment, dependence, identity, and meanings; however, because of the inherent nature of place-based work, these quantitative and qualitative questions often need to be tweaked to be site- or use-appropriate. This can make broader interpretation and application of the findings more difficult.

One example is the connection between recreationists and sense of place. A number of studies have found positive relationships between recreation and sense of place values. In several recreation studies (e.g. [25, 33–35]), recreationists did not have particularly high attachment to a place. Further, research that applies mixed qualitative and quantitative methods may help to tease out the reasons for these differences. A high sense of place also does not necessarily translate to actions. Rudestam [36] found a strong sense of place for waters in the Willamette River Basin among water users in their professional capacity (agriculture-related, fisheries-related, recreational outfitters), as well as those involved in agencies and watershed councils. However, that attachment did not motivate intentions for personal sacrifices. For example, interview participants still talked about clearcutting their lands for high timber prices or anglers being unhappy with management actions that would limit their catch. These results point to the importance of using sense of place values along with other social measures, as well as biophysical measures. Although social attitudes and values are complex and subjective, a better understanding of these attitudes and values, including sense of place, could enable better connections between communities and management and conservation of resources.

Applications of sense of place findings are primarily at a localized level but provide insights about the relative social value of different places. To increase the application of sense of place in environmental management, social data will need to be collected on a broader scale with more consistent means of data collection such as the use of standardized sense of place scales for use across locations. One promising advancement has been an increase in applying spatial techniques in place-based research (e.g., [37–39]). These techniques connect survey or interview data to places and landscapes which allows for the incorporation of spatially defined ecological data to analyze relationships with sense of place.

Beckley [4] calls for research that identifies the environmental attributes for which people develop place attachment; however, a big limitation in the use of sense of place, or many other place-based social indicators, for water quality assessment is the lack of localized biophysical water quality data. Place-based values are site specific and are not always generalizable past that location as testing and monitoring methods are not always consistent. Without corresponding localized water quality data—either perceptions or biophysical measurements—we may be able to capture the social value for that place as a whole but not for its environmental attributes. In the case of water quality valuation, this means that we are limited in our ability to explain changes in the social value for sense of place resulting from changes in water quality due to gaps in our biophysical monitoring and understanding.

5. Sense of place and environmental change

Sense of place is a social construction of place identity, dependence, and attachment that is mediated by physical attributes and conditions [4, 40]. For example, while many studies have found that long-term residents have higher place attachment, in Montana, McCool and Martin [41] surprisingly found that newer residents had higher place attachment. They explained this unusual finding as possibly reflecting the fact that many newer residents had moved to the area specifically because of the mountain access and environmental attributes of the area. Kibler et al. [42] highlighted the value of connecting human attachment to the condition of an ecosystem for evaluating the success of restoration projects. Specifically, they hypothesized that it is likely for ecosystem improvement in restoration projects to depend on the interaction between ecosystem function and sense of place. These interactions project that a restoration site where stakeholders have a high sense of place and where there is a highly functioning ecosystem will lead to emotionally invested stakeholders and iterative monitoring of the ecosystem. On the other end of the spectrum, they hypothesized that low sense of place and low ecosystem function would require enhancing stakeholder attachment for a restoration effort to be successful.

Minimal work has been conducted that moves beyond general attachment to the environment to directly connect sense of place to environmental attribute data. The connection of biophysical data with social data is often limited by the availability of the two types of data at the same meaningful scale. Many social scientists focus on survey respondents' or interview participants' environmental perceptions or landscape values for a location when biophysical data are unavailable [43]. For example, in Norway, Kaltenborn [44] found the most important contributing attribute to place attachment to be the perception of the quality of the natural environment. Brown and Raymond [37] investigated the relationships between landscape values which incorporate both ecological and social values, and sense of place in Australia. They found esthetic, spiritual, future generation, and wilderness values to be the best predictors of place attachment. Matarrita-Cascante et al. [45] found that natural amenities increased place attachment for both seasonal and permanent residents. Larson et al. [46] applied nine natural environmental wellbeing factors to explain sense of place values, including general "environmental quality," "water quality," "fishing," "soil," and others. They found that coastal residents valued beauty and conditions of the environment.

To date, very little work has connected sense of place with water quality assessments (see **Table 1**). In the only work that directly connects biophysical water quality data with sense-of-place meanings, Stedman [20] connected water quality to place attachment and satisfaction for property-owners in a lake-rich region of Wisconsin. He found that the construction of sense of place meanings was mediated by the level of shoreline development as well as the social influences of whether or not the lake felt like a wilderness escape place or a neighborhood of friends. In terms of place attachment, the two social influences (wilderness escape and sociability) that depend upon what the property owners were seeking essentially cancel one another out when considering shoreline development. More shoreline development leads to a more social environment, but also less wilderness, thereby differently affecting the experiences of each property owner. Jorgenson and Stedman [26] investigated the same dataset and found that perceptions of environmental features were the best predictors of place identity, dependence, and attachment.

There has been some work connecting various water quality metrics to sense of place without using biophysical data. Brehm et al. [5] measured predictors of water quality concern and found place meanings to be linked to local environmental concern. They found that the level of water quality concern was predicted by the environmental values, gender (female > male), and assigned place meanings (how impacted they perceived the watershed was by environmental threats and how they perceived the watershed as a getaway). Smith et al. [10] connected perceived ecological integrity (along with a set of other place attachment indicators) with a set of desired social and ecological outcomes for lakes in Illinois. They found that the more people believed the lakes contributed to the ecological integrity of the area, the more they desired improved environmental outcomes and the less they

Author	Water quality metric(s) used
Stedman [20]*	Level of lake shoreline development (number of structures within a 100 m buffer of the lake), water clarity, algal biomass, chlorophyll, color, alkalinity, and conductivity
Brehm et al. [5]	Water quality concern
Smith et al. [10]	Perceived ecological integrity Likert-scale questions: 1. This lake is important in protecting the landscape from development 2. This lake is important in providing habitat for wildlife 3. This lake is important in protecting water quality
Cox et al. [47]	Perceived waterway condition Likert-scale questions: 1. Considering everything, how would you rate the overall condition of the following waterways? 2. How would you rate the waterways in terms of the quality of the water? 3. How would you rate the waterways in terms of the vegetation along the shores? 4. How would you rate the waterways in terms of the number and variety of animals?
Larson et al. [46]	Environmental wellbeing factors for fishing, swimming, air quality, water quality, soil quality, beauty of the landscape, condition of the landscape, access to the natural areas, biodiversity, overall-natural environment

**This is the only study that applied specific water quality data. The other studies applied perceived ecological integrity or water quality value.*

Table 1.
Past research connecting water quality metrics and sense of place.

desired competing economic outcomes. Cox et al. [47] also investigated water quality perceptions and found a weak connection with the number of visits, which then indirectly affected other quality of life indicators, including sense of place. In a qualitative investigation of sense of place, Lukacs and Ardoin [12] found that the perception of the environmental attributes and biophysical resources influenced sense of place and watershed group participation in Appalachia.

6. Sense of place as a cultural ecosystem service indicator for water quality

Sense of place is sometimes identified as a cultural ecosystem service [48, 49]. The term "cultural ecosystem services" is used to represent a range of non-material benefits that humans receive from their interactions with the environment, including esthetic appreciation, spiritual services, cultural identity, recreation experiences and more [49, 50]. These types of benefits provide some convincing reasons for environmental protection that may be compelling to different audiences than those who prefer other ecosystem service benefits [51]. Because of the importance of cultural ecosystem service benefits to humans, their assessment is critical for understanding the impact of environmental change, including water quality degradation or improvement.

Although cultural ecosystem services are one of the core components of most ecosystem services frameworks, their assessment and use remains relatively limited [48, 51]. This is due, in part, to the challenge of calculating the economic value of the benefits, resulting in the frequent omission of the value of non-material cultural services [51]. Over the past few decades, a number of survey scales to capture sense of place attitudes for various geographies have been developed. As discussed throughout this chapter, very little of this work has been conducted in freshwater places and even less has been done in saltwater places.

Here, we propose a set of scaled sense of place questions (**Figure 2**) with the purpose of understanding the social impacts of water quality changes through recreation. These questions were compiled and modified from past work on sense of place in both water recreation areas as well as in other contexts. The questions are derived most directly from the work of Jorgensen and Stedman [26], Williams and Vaske [23], and Mullendore et al. [31]. In addition to the nine quantitative, Likert-scale questions, we also include open-ended, qualitative questions that can be used to further explain sense of place responses. We developed the scale questions and the qualitative follow-up questions to attempt to address some of the issues mentioned above. The scale is intended for increased consistency in data collection and an increased ability to compare sense of place across different geographic places. The qualitative questions are intended to capture some of the nuance associated with the complexity of sense of place and to better capture place meanings. In future work, we will explore the use of our sense of place scale to further elucidate variations in economic values for changes in water quality, an area that has not been explored by researchers to date.

To develop the quantitative and qualitative questions, we began with a set of open-ended qualitative questions gleaned from the past sense of place research. We then modified a number of these questions and, through further focus group testing, reduced the set to nine questions capturing the three subcomponents of sense of place. The two qualitative questions were also refined through focus group testing.

As discussed earlier in this chapter, scaled sense of place questions have been used in a range of different research efforts. These include connecting sense of place

Thinking about the place you visited most recently for water recreation, please indicate how much you agree or disagree with the following statements. (Fill in one circle for each statement)

		Strongly Disagree	⟵		⟷		⟶	Strongly Agree
Place Dependence	That place provides value to me that I can't obtain elsewhere.	O	O	O	O	O	O	O
	I get more satisfaction out of visiting that place than any other recreation place.	O	O	O	O	O	O	O
Place Identity	The recreational activities that I pursue at that place say a lot about who I am.	O	O	O	O	O	O	O
	I identify with the physical landscape of that place.	O	O	O	O	O	O	O
Place Attachment	That place means a lot to me.	O	O	O	O	O	O	O
	I have no particular love for that place compared to other areas.	O	O	O	O	O	O	O
	I am very attached to that place.	O	O	O	O	O	O	O
	I feel happiest when I'm at that place.	O	O	O	O	O	O	O
	Many important memories are tied to that place.	O	O	O	O	O	O	O

Why did you choose that place? (Please describe)

What is important to you about that place, if anything?

Figure 2.
Sense of place scale and qualitative questions developed for use in water quality social assessments in the context of coastal recreation.

with support for environmental actions, recreation behaviors, and perceived environmental quality. Our scaled questions provide specific metrics for quantifying place dependence, identity, and attachment with the ability to use the data to better understand the impacts of changes in water quality at recreational sites. By coupling the sense of place data with biophysical data, we will be able to conduct analyses of how sense of place is affected by water quality. These analyses will connect site-level water quality data such as water clarity via Secchi depth measurements, bacteria counts from beach monitoring, or chlorophyll *a* to the sense of place measurements, like in the work of Stedman [20] in Wisconsin lakes.

In the past, a great deal of the sense of place research has focused on qualitative exploration of the concepts of place attachment, identity, and dependence (e.g., [15, 32, 40]). The symbolic and complex meaning of sense of place and its components makes agreement/disagreement with simplified statements, like those required to develop a scale, difficult to capture in their entirety [5]. Qualitative questioning, including the two qualitative questions proposed in **Figure 2** ("Why did you choose that place? (Please describe)" and "What is important to you about that place, if anything?") allows for deeper exploration of the meaning behind the responses given in the quantitative scale and an extension of sense of place meanings.

7. Conclusion

This work expands on the research investigating the relationships between biophysical data and social data, specifically in the context of evaluating the relationships between sense of place and water quality. Davenport and Anderson [15] wrote "A holistic and integrated understanding is needed, though, of place meanings and the setting to which these meanings are ascribed. What happens to sense of place when places change? What happens when landscape change threatens place meanings and emotions? (p. 630)" Although a number of researchers have contributed to the sense of place literature since then, direct investigations of the impacts on sense of place from changes to the environment remain relatively non-existent.

We have presented a set of sense of place scales that capture the three main components of sense of place – place dependence, identity, and attachment. We combine these scales with qualitative questions in order to further understand the nuance of people's sense of place. Through our work, we are attempting to advance the research on sense of place, as well as contribute to better understanding social values for water quality. Used in conjunction with environmental economic valuation methods for recreation and water quality, sense of place may provide additional nuance and explanatory power in describing people's preferences for the quality of natural resources. We suggest that this approach may be useful in other places and contexts.

Sense of place is a promising metric in the assessment of water quality for capturing the social value of various locations. Moving forward, in order to identify the impacts of changes in water quality and better inform the process of managing resources, increased social and biophysical data are needed at place-based scales. If researchers collect and report these data in more consistent ways across places, it will be possible to make comparisons across places and contexts. Increased collection and application of place-based social data can contribute to understanding community priorities for conservation or restoration, which is crucial for informing targeted management aimed at water quality improvements. Identifying areas of particular value may also help to identify potential sources of conflict or areas of special value. By informing water quality management to better target waters, community priorities may be better accounted for in interventions and decisions. Finally, sense of place research can also be used to improve connections between humans and natural systems by understanding the social and environmental attributes that make a place important.

Author details

Kate K. Mulvaney*, Nathaniel H. Merrill and Marisa J. Mazzotta
U.S. Environmental Protection Agency, Office of Research and Development, Center for Environmental Measurement and Modeling, Atlantic Coastal Environmental Sciences Division, Narragansett, Rhode Island, USA

*Address all correspondence to: mulvaney.kate@epa.gov

References

[1] Bonaiuto M, Carrus G, Martorella H, Bonnes M. Local identity processes and environmental attitudes in land use changes: The case of natural protected areas. Journal of Economic Psychology. 2002;**23**:631-653

[2] Farnum J, Hall T, Kruger LE. Sense of place in natural resource recreation and tourism: An evaluation and assessment of research findings. United States Department of Agriculture General Technical Report PNW-GTR-660; 2005

[3] U.S. EPA (U.S Environmental Protection Agency). Community Culture and the Environment: A Guide to Understanding a Sense of Place. U.S. EPA Report (EPA 842-B-01-003); 2002

[4] Beckley TM. The relative importance of sociocultural and ecological factors in attachment to place. In: Krueger L, editor. Understanding Community-Forest Relations. U.S. Forest Service. General Technical Report PNW-GTR-566; 2003

[5] Brehm JM, Eisenhauer BW, Stedman RC. Environmental concern: Examining the role of place meaning and place attachment. Society and Natural Resources. 2013;**26**(5):522-538

[6] Halpenny EA. Pro-environmental behaviors and park visitors: The effect of place attachment. Journal of Environmental Psychology. 2010;**30**:409-421

[7] Kaltenborn BP, Williams DR. The meaning of place: Attachments to Femundsmarka National Park, Norway, among tourists and locals. Norwegian Journal of Geography. 2002;**56**:189-198

[8] Mesch GS, Manor O. Social ties, environmental perceptions, and local attachment. Environment and Behavior. 1998;**30**(4):504-519

[9] Ramkisson H, Smith LDG, Weiler B. Testing the dimensionality of place attachment and its relationships with place satisfaction and pro-environmental behaviours: A structural equation approach. Tourism Management. 2013;**36**:552-566

[10] Smith JW, Davenport MA, Anderson DH, Leahy JE. Place meanings and desired management outcomes. Landscape and Urban Planning. 2011;**101**:359-370

[11] Vaske JJ, Kobrin KC. Place attachment and environmentally responsible behavior. The Journal of Environmental Education. 2001;**32**(4):16-21

[12] Lukacs HA, Ardoin NM. The relationship of place re-making and watershed group participation in Appalachia. Society and Natural Resources. 2014;**27**(1):55-69

[13] Devine-Wright P. Rethinking NIMBYism: The role of place attachment and place identity in explaining place-protective action. Journal of Community and Applied Social Psychology. 2009;**19**:426-441

[14] Williams DR, Stewart SI. Sense of place: An elusive concept that is finding a home in ecosystem management. Journal of Forestry. 1998;**96**(5):18-23

[15] Davenport MA, Anderson DH. Getting from sense of place to place-based management: An interpretive investigation of place meanings and perceptions of landscape change. Society and Natural Resources. 2005;**18**(7):625-641

[16] Manzo LC. Beyond house and haven: Toward a revisioning of emotional relationships with places. Journal of Environmental Psychology. 2003;**23**:47-61

[17] Poe MR, Donatuto J, Satterfield T. "Sense of place": Human wellbeing considerations for ecological restoration in Puget Sound. Coastal Management. 2016;**44**(5):409-426

[18] Stedman RC, Ingalls M. C. 10 Topophilia, biophilia, and greening in the red zone. In: Tidball KG, Krasny MS, editors. Greening in the Red Zone: Disaster. Dordrecht: Resilience and Community Greening. Springer; 2014

[19] Campbell C. "Our dear north country": Regional identity and national meaning in Ontario's Georgian bay. Journal of Canadian Studies. 2002-2003;**37**(4):68-91

[20] Stedman RC. Is it really just a social construction? The contribution of the physical environment to sense of place. Society & Natural Resources. 2003;**16**:67-685

[21] Twigger-Ross CL, Uzzell DL. Place and identity processes. Journal of Environmental Psychology. 1996;**16**:205-220

[22] Williams DR, Patterson ME, Roggenbuck JW. Beyond the community metaphor: Examining emotional and symbolic attachment to place. Leisure Sciences. 1992;**14**:29-46

[23] Williams DR, Vaske JJ. The measurement of place attachment: Validity and generalizability of a psychometric approach. Forest Science. 2003;**49**(6):830-840

[24] Anton CE, Lawrence C. Home is where the heart is: The effect of place of residence on place attachment and community participation. Journal of Environmental Psychology. 2014;**40**:451-461

[25] Bricker KS, Kerstetter DL. Level of specialization and place attachment: An exploratory study of whitewater recreationists. Leisure Sciences. 2000;**22**:233-257

[26] Jorgenson BS, Stedman RC. A comparative analysis of predictors of sense of place dimensions: Attachment to, dependence on, and identification with lakeshore properties. Journal of Environmental Management. 2006;**79**:316-327

[27] Scannell L, Gifford R. Defining place attachment: A tripartite organizing framework. Journal of Environmental Psychology. 2010;**30**:1-10

[28] Tuan YF. Topophilia: A Study of Environmental Perception, Attitudes, and Values. Englewood Cliffs, NJ, USA: Prentice Hall; 1974

[29] Hammitt WE, Backlund EA, Bixler RD. Experience use history, place bonding and resource substitution of trout anglers during recreation engagements. Journal of Leisure Research. 2004;**36**(3):356-378

[30] Simoni J, Floress K. An exploration of place meanings among residents of Central Wisconsin. Lake and Reservoir Management. 2015;**31**(1):1-10

[31] Mullendore ND, Ulrich-Schad JD, Prokopy LS. U.S. farmers' sense of place and its relation to conservation behavior. Landscape and Urban Planning. 2015;**140**:67-75

[32] Jacobs MH, Buijs AE. Understanding stakeholders' attitudes toward water management interventions: Role of place meanings. Water Resources Research. 2011;**47**:1

[33] Kyle G, Graefe A, Manning R, Baron J. Effects of place attachment on users' perceptions of social and environmental conditions in a natural setting. Journal of Environmental Psychology. 2004;**24**:213-225

[34] Moore RL, Graefe AR. Attachments to recreation settings: The case of rail-trail users. Leisure Sciences. 1994;**16**(1):17-31

[35] White DD, Virden RJ, van Riper CJ. Effects of place identity, place dependence, and experience-use history on perceptions of recreation impacts in a natural setting. Environmental Management. 2008;**42**(4):647-657

[36] Rudestam K. Loving water, resenting regulation: Sense of place and water management in the Willamette watershed. Society and Natural Resources. 2014;**27**(1):20-35

[37] Brown G, Raymond C. The relationship between place attachment and landscape values: Toward mapping place attachment. Applied Geography. 2007;**27**:89-111

[38] McLain R, Poe M, Biedenweg K, Cerveny L, Besser D, Blahna D. Making sense of human ecology mapping: An overview of approaches to integrating socio-spatial data into environmental planning. 2013;**41**:651-665

[39] Van Riper CJ, Kyle GT, Sutton SG, Barnes M, Sherrouse BC. Mapping outdoor recreationists' perceived social values for ecosystem services at Hinchinbrook Island National Park, Australia. Applied Geography. 2014;**35**:164-173

[40] Sampson KA, Goodrich CG. Making place: Identity construction and community formation through 'sense of place' in Westland, New Zealand. Society and Natural Resources. 2009;**22**(10):901-915

[41] McCool S, Martin S. Community attachment and attitudes toward tourism development. Journal of Travel Research. 1994;**32**(3):29-34

[42] Kibler KM, Cook GS, Chambers LG, Donnelly M, Hawthorne TL, Rivera FJ, et al. Integrating sense of place into ecosystem restoration: A novel approach to achieve synergistic social-ecological impact. Ecology and Society. 2018;**23**(4):25

[43] Brown G. Mapping spatial attributes in survey research for natural resource management: Methods and applications. Society and Natural Resources. 2005;**18**:17-39

[44] Kaltenborn BP. B.P. nature of place attachment: A study among recreation homeowners in Southern Norway. Leisure Sciences. 1997;**19**(3):175-189

[45] Matarrita-Cascante D, Stedman R, Luloff AE. Permanent and seasonal residents' community attachment in natural amenity-rich areas. Environment and Behavior. 2010;**42**(2):197-220

[46] Larson S, De Freitas DM, Hicks CC. Sense of place as a determinant of people's attitudes towards the environment: Implications for natural resources management and planning in the great barrier reef, Australia. Journal of Environmental Management. 2013;**117**:226-234

[47] Cox ME, Johnstone R, Robinson J. Relationships between perceived coastal waterway condition and social aspects of quality of life. Ecology and Society. 2006;**11**(1):35

[48] Hernández-Morcillo M, Plieninger T, Bieling C. An Empirical Review of Cultural Ecosystem Service Indicators. Ecological Indicators. 2013;**29**:434-444

[49] Millennium Ecosystem Assessment. Ecosystems and Human Well-Being: Synthesis. Washington, DC: Island Press; 2005

[50] Urquhart J, Acott T. A sense of place in cultural ecosystem services: The case of Cornish fishing communities. Society and Natural Resources. 2014;**27**(1):3-19

[51] Chan KMA, Satterfield T, Goldstein J. Rethinking ecosystem services to better address and navigate cultural values. Ecological Economics. 2012;**74**:8-18

Section 2

Water Quality Treatments

Chapter 3

Natural Wastewater Treatment Systems for Prevention and Control of Soil-Transmitted Helminths

Abdallah Zacharia, Anne H. Outwater and Rob Van Deun

Abstract

Wastewater reuse has been considered as an alternative way of overcoming water scarcity in many parts of the world. However, exposures to wastewater are associated with higher prevalence of soil-transmitted helminths (STHs). Globally, about two billion people are infected with at least one species of STHs with those having heavy infections presenting considerable morbidities. The most serious STH species infecting humans include roundworm (*Ascaris lumbricoides*), whip-worm (*Trichuris trichiura*), and hookworms (*Necator americanus* and *Ancylostoma duodenale*). Despite ongoing control campaigns using preventive chemotherapy, wastewater in endemic countries still contains concentrations of STH eggs that put exposed populations at risk of infection. According to the World Health Organization, we can achieve sustainable control of STH by using improved sanitation systems. Since natural wastewater treatment systems (waste stabilization ponds and constructed wetlands) require low maintenance and operational costs, have low mechanical technology and energy consumption, they are ideal for sustainable sanitation services. In addition, natural wastewater treatment systems are reported to efficiently remove various pathogenic organisms from wastewater. This chapter explains the role of natural wastewater treatment systems as sustainable sanitation facilities in removing STH from wastewater and therefore preventing disease transmission.

Keywords: *Ascaris lumbricoides*, constructed wetlands, hookworms, soil-transmitted helminths, *Trichuris trichiura*, wastewater reuse, waste stabilization ponds

1. Introduction

Population growth significantly contributes to water shortages in about 100 countries worldwide. It is estimated that by the year 2025, two-thirds of all people will be experiencing moderate to severe fresh water shortage [1]. Wastewater reuse has been considered as an alternative way of overcoming water scarcity in various parts of the world [2]. Treated and untreated wastewaters have been applied to economic and domestic activities including industry (applied in cooling and cleaning); recreation (swimming pools, irrigation of parks, and golf courses); and agriculture (irrigation) [3]. Globally more than 20 million hectares of agricultural land is irrigated with either treated or untreated wastewater [4]. In addition to the direct uses, about 80% of all wastewater is discharged to the world's water bodies such as rivers, lakes, swamps, and streams [5].

Whether it is used directly or indirectly, an important consideration in wastewater reuse is its quality in terms of pollutant types and content. Wastewater reuse or discharge to surface water poses risks of disease transmission from animal and/or human-excreted waterborne pathogenic organisms to exposed communities [6, 7]. Transmissions of pathogenic bacteria are frequently a public health concern; however, the most important public health problem is parasite transmission [8]. Among the pathogenic parasites identified in wastewater, soil-transmitted helminths (STHs) are the most common. The problem of STH predominance in wastewater is measured in terms of how frequently the parasites are identified and their level of concentration [9]. The predominance of STHs in wastewater has been associated with the ability of their eggs to resist different types of environmental conditions compared to other organisms [10, 11].

In many countries, exposure to wastewater has been associated with high prevalence of STH transmission [12–14]. In addition, STHs are more prevalent in low- and middle-income countries where more than 72% of generated wastewater is discharged without being treated [15, 16]. To prevent the spread of helminthic diseases such as those caused by STH (e.g., ascariasis, trichuriasis, and hookworm), several measures to protect health have been practiced in wastewater reuse. These measures include wastewater treatment, crop restrictions, control of wastewater application, control of human exposure, and promotion of personal hygiene. Of these measures, wastewater treatment is the most commonly adopted approach in many controlled wastewater reuse schemes [17]. **Figure 1** presents an estimation of wastewater treatment capacities in 2015 in countries classified by level of income and their expected achievement by 2030. The estimation in 2015 shows that the capacity of wastewater treatment is 70% of all wastewater generated in high-income countries and 8% of all wastewater generated in low-income countries [5].

Compared to conventional treatment systems such as activated sludge and trickling filters, natural wastewater treatment systems have been reported to be more efficient at removing STH eggs from wastewater [18]. The potential of two types of natural wastewater treatment systems (waste stabilization ponds and constructed wetlands) for prevention of STH infections is discussed in this chapter.

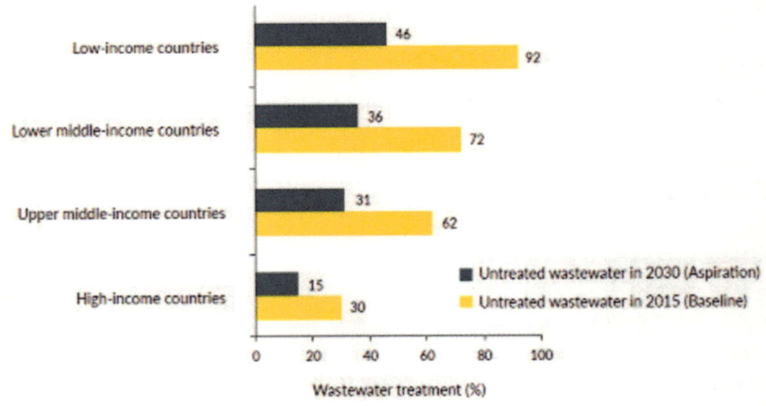

Figure 1.
Wastewater treatment in countries as classified based on the level of income. Figure from [16].

2. Soil-transmitted helminths in human and wastewater

2.1 Soil-transmitted helminths in humans

STHs, are also known as geohelminths, are multicellular intestinal nematodes. Part of their life cycle depends on soil for maturation and they are transmitted through contaminated soil. The important STH species infecting humans include roundworm (*Ascaris lumbricoides*), whipworm (*Trichuris trichiura*), and hookworms (*Necator americanus and Ancylostoma duodenale).* These helminths are distributed throughout the world. Globally, about two billion people are infected with at least one species of STHs with those having heavy infections presenting considerable morbidities including malnutrition, allergy, and respiratory difficulties including asthma and Löffler's syndrome, diarrhea, intestinal obstruction, rectal prolapse, anemia, and cognitive development problems [19]. With limited access to clean and safe water often leading to poor hygiene and insufficient sanitation services, frequency of helminthiasis is higher in low- and middle-income countries than in high-income countries [15].

2.2 Life cycles of soil-transmitted helminths

Transmission of STH occurs through the fecal-oral route by ingesting viable eggs of *Ascaris lumbricoides* and *Trichuris trichiura* from contaminated soil or food or, through skin penetration by third-stage hookworm larvae (filariform larvae). Based on the passage of young-stage worms (larvae), the STHs are divided into three groups. *Trichuris trichiura* undergoes a direct life cycle whereby the ingested eggs directly develop to adult worms inside human intestines. *Ascaris lumbricoides* undergoes a so-called modified direct life cycle whereby ingested eggs hatch to release larvae in the human intestine. The released larvae penetrate intestinal mucosa to the blood stream where they migrate to the liver, heart, lung, upper respiratory track, then return to the intestines where they develop into adults. Unlike *Ascaris lumbricoides* and *Trichuris trichiura,* eggs of hookworms hatch in the soil where they develop to the infective stage-three larvae (filariform larvae). The filariform larvae penetrate unbroken skin of human beings to the blood, and migrate to the liver, heart, lung, upper respiratory tract to the intestine where they mature to adults (**Figure 2**). In very rare cases, hookworm transmission occurs via the fecal-oral route. In the intestine, the sexually mature male and female adults mate and the female lays fertile eggs. In all these helminth species, eggs are excreted with feces to the environment. When they reach the soil, they mature and become infective (*Ascaris lumbricoides* and *Trichuris trichiura*), or hatch to rhabditiform larvae, which then develop into the infective filariform larvae (hookworms). STHs do not multiply in the host. Therefore, each one that is found in the intestine is the result of a single infection event [20].

2.3 Soil-transmitted helminth treatment and control

The drugs of choice for treatment of STHs are albendazole (400 mg) and mebendazole (500 mg). Measures used to control STH involve periodic deworming of at-risk groups to eliminate infective worms, health education to prevent infection and reinfection, and improved sanitation to reduce soil contamination with infective eggs.

In 2011, the World Health Organization (WHO) opted for the use of periodic mass treatment with albendazole for at-risk people in STH endemic areas.

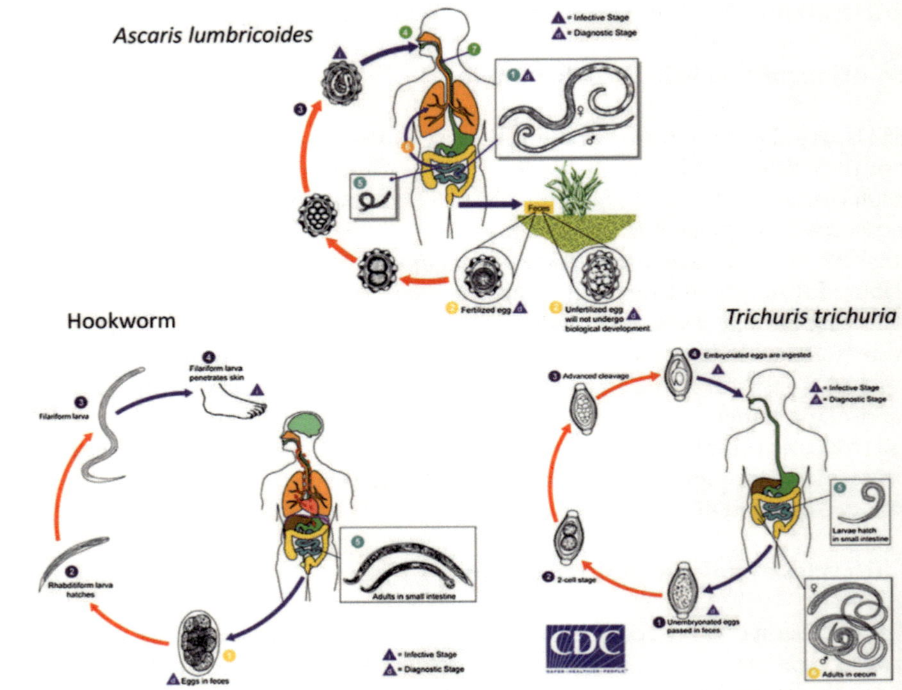

Figure 2.
Life cycles of three species of soil-transmitted helminths. Adapted from CDC, creative commons (Center for Disease Control and Prevention, 2015. Parasites [online]. Available at: http://www.cdc.gov/parasites/).

The objective was to control morbidity by reaching 75% coverage of preschool- and school-age children by the year 2020. Based on data collected in 2018, 68% of preschool-age children and 73% of school-age children in endemic countries who were in need of treatment received it during the 8 years of implementation (between 2010 and 2017). In spite of this achievement, transmission still continues.

In 2018, the WHO set six targeted for STH control programs in the period of 2020–2030. These targets include: achieving and sustaining elimination of STH morbidities in preschool- and school-age children by 2030, reducing the number of anthelminthic tablets required for preventive chemotherapy (PC), increasing financial support in endemic countries by their own governments for PC, establishing an efficient STH control program to women of reproductive age, establishing an efficient strongyloidiasis control program for school-age children, and ensuring universal access to at least basic sanitation and hygiene by 2030 in STH endemic areas [21].

2.4 Soil-transmitted helminths in wastewater

STHs are among the most frequently identified pathogens in wastewater. STHs may enter wastewater ways from both point and nonpoint sources. Domestic wastewater, by definition, is always contaminated with human and animal excreta. STHs are introduced into wastewater through direct discharge of human excreta (feces) containing eggs. STH can also enter wastewater through discharge of sewage to water ways and through water/rain runoff from contaminated soil (where humans practice open defecation) or agricultural lands using human and animal excreta as manure.

Commonly, STHs in wastewater occur in the form of eggs. Eggs are the most environmentally resistant stage of STH. They can persist outside of their host bodies for up to 9 months [11]. STH eggs contain several shells, three to four layers depending on the genera. These shells are made up of lipoprotein and protein structures laminating the egg cell providing resistance against external physicochemical stresses. A thick outer layer gives the egg protection. The middle layer consists of several sub-layers and is important for prevention against physical destruction as well as giving the egg its shape. The inner layer is preventive against fatal chemicals such as strong acids, bases, oxidants and reducing agents, detergents, and proteolytic compounds. It also protects the egg from desiccation. Alongside the stated functions, these layers allow for gaseous exchange and water passage [22].

The resistant nature of STH eggs allows them to remain viable in external environments such as wastewater. For example, *Ascaris lumbricoides* eggs have been found to be as viable in wastewater as in fresh stool samples. In addition, the eggs embryonate after being exposed to aerobic conditions. Hookworm eggs remain viable in anaerobic conditions for up to 2 weeks. They hatch when they are in an aerobic condition to release first-stage (rhabditiform) larva. However, the released larvae seem unable to develop to the infective stage-L3 (filariform) larva [11].

Table 1 presents concentrations of different species of STHs in raw wastewater or wastewater sludge reported in various STH endemic countries with preventive chemotherapy intervention campaigns. These findings are from research conducted a short time before the start of PC, within, or after a PC campaign (data collected between 2009 and 2018). Studies conducted between the year 2014 and 2018 in India, Lesotho, Malawi, South Africa, Tanzania, and Cameroon recorded high concentrations of STH eggs in either wastewater and/or sludge samples [24–28, 31, 33]. These countries had less than 5 years of PC implementation with coverage of more than 75% by the year 2017 [21]. Despite being in a group of few countries that have controlled moderate and heavy intensity of STH infection to less than 1% [21], a study conducted in Senegal in 2016 presented high concentration of STH eggs in sludge samples collected from wastewater treatment plants [30]. The presence of high concentrations of STH eggs in wastewater and sludge in countries with high coverage of PC implementation could be attributed to the fact that the campaign is selective for some at-risk groups (preschool- and school-age children) and leaves out others such as adults working in high-risk areas. These findings demonstrate that the risk of STH transmission still exists especially in communities exposed to wastewater, and wastewater-produced products such as vegetables. This solicits for the need of interventions that will prevent transmission and bring effects across all at-risk groups.

Concentrations of STH eggs in wastewater vary from one country to another. Variations also exist within different parts of the same country and even between sampling points (**Table 1**). Several reasons may account for the variation in concentration of STH eggs in wastewater. Factors include: endemicity of the area's source of wastewater, volume of wastewater sampled, and diagnostic methods used. In areas with high STH endemicity, the concentration of eggs in wastewater is expected to be higher compared to low endemicity areas. The sources of wastewater affect the concentration of eggs since wastewater collected directly from toilets, latrines, or septic tanks contains high concentrations of fecal matter compared to that collected from other sources such as wastewater treatment systems or contaminated rivers. When domestic wastewater is mixed with wastewater from other sources (industrial or rain runoff) in the treatment systems or rivers, dilution of fecal contents (including STH eggs) in domestic wastewater occurs and therefore lowers its concentration. This is clearly depicted in **Table 1**, whereby in many cases concentrations of STH eggs were higher in wastewater and sludge collected from latrines and trucks compared to that collected from influents of treatment systems. The volume and type of diagnostic method have an influence on the determined

Country	Source	STH species	Mean(s) eggs/L or g	Reference
Burkina Faso	Wastewater from WWTP	*A. lumbricoides*	7	[23]
		Hookworms	6	
		T. trichiura	1.40	
Cameroon	Wastewater from marshy areas	*A. lumbricoides*	77.75	[24]
		Hookworms	59	
		T. trichiura	115.67	
	Sludge from latrines	*A. lumbricoides*	16667[a]	[25]
		Hookworms	16611[a]	
		T. trichiura	13444[a]	
Ghana	Wastewater from farm	*A. lumbricoides*	2.72	[14]
		Hookworms	1.72	
Lesotho	Wastewater from WWTP	*A. lumbricoides*	87	[26]
		Hookworms	26	
		T. trichiura	12	
Malawi	Sludge from pit latrines	*A. lumbricoides*	0.4 and 4.7[*]	[27, 28]
		Hookworms	7.65 and 20.5[*]	
		T. trichiura	0.06[*]	
Nigeria	Wastewater from WWTP	*A. lumbricoides*	307	[29]
		Hookworms	135	
		T. trichiura	92	
Senegal	Sludge from WWTP	*A. lumbricoides*	1079[*]	[30]
		Hookworms	257[*]	
		T. trichiura	1647[*]	
South Africa	Wastewater from WWTP	*A. lumbricoides*	54	[26]
		Hookworms	31.33	
		T. trichiura	14.53	
	Sludge from WWTP	*A. lumbricoides*	722[*]	[30]
		Hookworms	334[*]	
		T. trichiura	154[*]	
Tanzania	Wastewater from WWTP	*A. lumbricoides*	13.67	[31]
		Hookworms	20.75	
		T. trichiura	20	
Uganda	Wastewater from channel	*A. lumbricoides*	4	[32]
		Hookworms	27	
India	Wastewater from shared toilet	*A. lumbricoides*	58	[33]
		Hookworms	25,174	
		T. trichiura	38	
Indonesia	Wastewater from trucks and farm	*A. lumbricoides*	18.24 and 119.44	[34, 35]
		Hookworms	51.29	

Country	Source	STH species	Mean(s) eggs/L or g	Reference
Brazil	Wastewater from WWTP	*Ascaris species* and Hookworms	300	[36]
Bolivia	Wastewater from WWTP	*A. lumbricoides*	324.33	[37]
		Hookworms	5.13	
		T. trichiura	29.02	
Colombia	Wastewater from WWTP	*A. lumbricoides*	72	[38]
		T. trichiura	1.60	
Peru	Wastewater from WWTP	*A. lumbricoides*	142	[39]

STH—Soil-transmitted helminth; WWTP—Wastewater treatment plant.
[a]*Median.*
[*]*Concentrations in sludge in eggs per gram.*

Table 1.
Soil-transmitted helminth concentrations reported in wastewater and sludge in various endemic countries with Albendazole preventive chemotherapy intervention implementation campaign.

concentration of STH eggs in wastewater. The larger the volume of wastewater, the higher the chance of STH eggs recovery and concentration. Also when the diagnostic method used had high eggs recovery efficiency, the chance of STH recovery increased as along with concentration [11].

3. Natural wastewater treatment systems

Natural wastewater treatment systems are biological treatment systems that require no or very little electrical energy; instead, they rely on entirely natural factors such as sunlight, temperature, filtration, adsorption, sedimentation, biodegradation, etc., to treat wastewater or fecal sludge. They utilize naturally occurring physicochemical and ecological processes in removing pollutants from wastewater. The processes involve interactions of microorganisms, aquatic plants, substrates (media), solar energy (temperature and light), and wind. These processes are important for removal of both physicochemical pollutants and biological (pathogenic) pollutants. Natural wastewater treatment systems have low maintenance and operational costs, low energy consumption, and low mechanical technology and are therefore ideal for sustainable sanitation services, especially in low- and middle-income countries [40]. Waste stabilization ponds (WSPs) and constructed wetlands (CWs) are common natural wastewater treatment systems used for treating wastewater from both point and nonpoint sources. They can be applied as a single standing treatment system or coupled with other treatment system(s). When used as part of larger treatment plants, they may be applied as primary, secondary, or tertiary systems. These systems are capable of efficiently removing varieties of wastewater pollutants including organic matter, nutrients, harmful chemicals, as well as pathogens [41]. Since the main purpose of this chapter is to provide information on the role played by natural wastewater treatment systems on prevention of STH, the following discussion focuses on mechanisms for their removal by these systems.

3.1 Waste stabilization ponds

WSPs are human-made shallow basins comprised of a single or series of anaerobic, facultative, or maturation ponds (**Figure 3**). They are used in either centralized

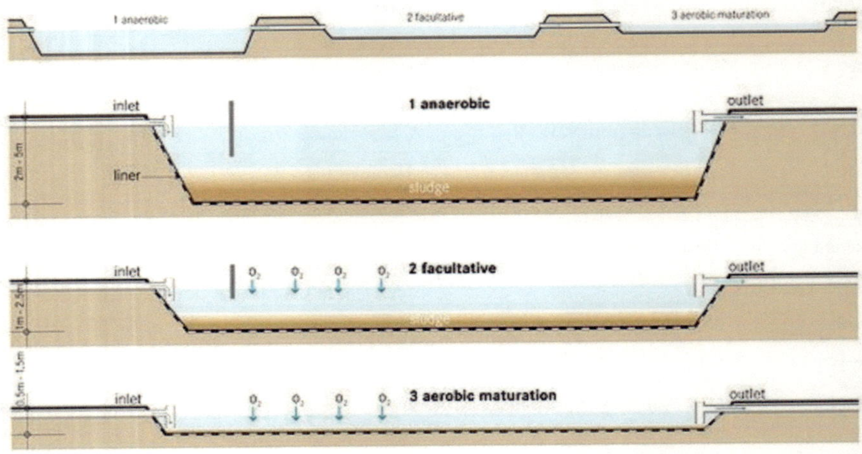

Figure 3.
Layout of typical waste stabilization pond system showing design of all three treatment stages.

or semi-centralized wastewater treatment plants serving connected households in towns and cities. Anaerobic ponds are used as pre-treatment. This part of the WSP system receives high organic loads of raw wastewater, often including septic tank sludge. The high organic loads produce anaerobic conditions throughout the pond. Anaerobic ponds are designed to remove particles (organic matter) through sedimentation or biological degradation. Facultative ponds are used as a secondary stage. Both aerobic and anaerobic processes occur in facultative ponds. Remaining biodegradable organic matter from anaerobic pond is removed in facultative pond, through the coordinated activity of algae and heterotrophic bacteria. Maturation pond is used as tertiary treatment before discharge to the outside environment. Their main function is the removal of pathogens. These ponds entirely use aerobic processes [42].

3.2 Constructed wetlands

Constructed wetlands (CWs) are human-made systems designed to utilize naturally occurring processes similar to those occurring in natural wetlands but in a controlled environment, for wastewater purification [43]. They consist of a bed of media (soil, gravel substrate) or liner and wetland plants (free floating, rooted emergent, or submerged). CWs are classified based on the position of the water surface (level) in relation to the surface of the soil or substrate; they can be surface flow (free water) or subsurface flow (**Figure 4**). In a surface flow, CW water level is positioned above the substrate and covered with wetland plants. This type of CW can be further classified based on the growth form of dominating vegetation as free floating, floating leafed, emergent, or submerged macrophytes. In a subsurface flow CW, wastewater is flowing through the porous media; the water level is positioned beneath the surface of the wetland media. This type of CW makes the use of emergent macrophytes only. The subsurface flow CW is further classified based on the predominant water flow direction in the system as horizontal or vertical. In horizontal subsurface flow CW, the predominant water flow direction is horizontal to the surface of the system while in vertical subsurface flow CW the predominant water flow direction is vertical to the surface of the system [42].

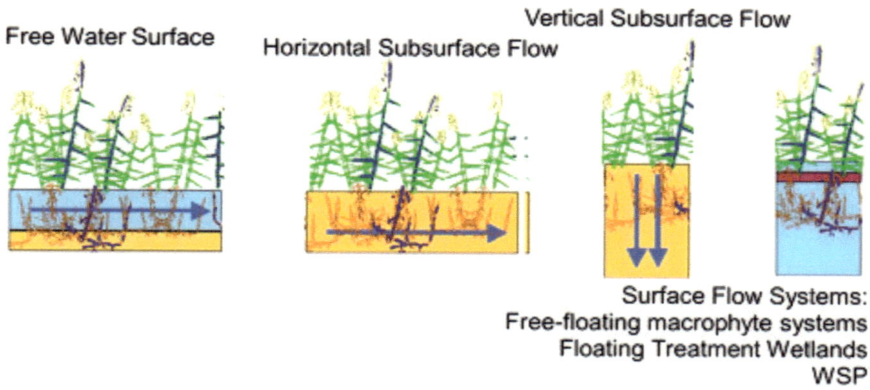

Figure 4.
Types of constructed wetlands systems. Picture from [42].

4. Soil-transmitted helminth removal in natural treatment systems

An actual risk of soil-transmitted helminthiasis to public health occurs when four conditions are present during wastewater reuse: (1) an infective dose of the helminths eggs reaches the field, (2) the infective dose reaches the human host, (3) the host becomes infected, and (4) the infection causes diseases or further transmission. If the first three conditions are present and the fourth is absent, the risk is just a potential risk. The WHO has set the health-based targets that can be used to reduce public health risk of helminths transmission. According to the WHO Guideline, helminth transmission among a wastewater-exposed population should not happen when wastewater quality is ≤1 helminth egg per liter. To achieve the set health-based target, a combination of health protection measures targeted at different areas of intervention should be implemented. The health protection measures include: (1) wastewater treatment or (2) a combination of wastewater treatment and thoroughly washing wastewater-irrigated produce to protect consumers, or (3) a combination of wastewater treatment and protection of workers by giving them personal protective equipment such as shoes and gloves. When children less than 15 years are part of an exposed population, extra measures are required. The extra measures include more stringent wastewater treatment in order to achieve wastewater quality of ≤0.1 helminth egg per liter, or providing PC with anthelminthic [44]. The above explanations show that wastewater treatments play a vital role in preventing STH transmission among exposed communities.

STH eggs cannot be inactivated by wastewater disinfection methods such as chlorine, ozone, temperature (unless above 40°C), or UV light applied in conventional systems because of their highly resistant nature caused by the three outer layers. Natural wastewater treatment systems are considered more effective at removing STH eggs from wastewater compared to conventional treatment systems such as activated sludge and trickling filters. Large sizes and high densities of most STH eggs allow them to be easily removed by mechanisms occurring in natural wastewater treatment systems (sedimentation, filtration, and adsorption). Natural wastewater treatment systems can remove 100% of helminth eggs from wastewater while conventional wastewater treatment processes can remove up to 90–99% of helminth eggs [45]. The higher efficiency of helminth egg removal by natural wastewater treatment systems prevents them from reaching the field or the exposed human hosts. Different types and designs of natural wastewater treatment systems have different helminth egg removal mechanisms and hence different efficiencies.

The commonly known natural wastewater treatment systems include WSP and CW. Studies conducted in different counties have shown that WSP systems are able to remove all STH eggs from wastewater. These systems have been shown to be efficient at removing helminth eggs in tropical countries like Burkina Faso, Honduras, Tanzania, Kenya, Bolivia, Brazil, and Colombia. They were also efficient in temperate countries as recorded in Iran, Morocco, Egypt, and Spain. However, sometimes WSP effluents have reported higher concentrations of STH eggs than that recommended by the WHO. Two out of five assessed WSP systems in Tunisia gave out effluents with more than one *Ascaris lumbricoides* eggs per liter [46], while one WSP in Tanzania and one WSP in Cayman Islands generated effluents with more than one hookworm eggs per liter [31, 47].

CW systems have also been shown to efficiently remove STH eggs from wastewater. The systems were observed to be more efficient when coupled with other treatment systems such as WSP [48]. Data collected from the few studies conducted to assess parasite removal efficiency of CW systems showed that, regardless of the influent concentration, this type of natural wastewater treatment could reduce the STH eggs to <1 per liter (**Table 2**).

Sedimentation is believed to be the primary removal mechanism in WSP and free water surface flow CW treatment systems. In subsurface flow CW, mechanical filtration and adsorption are the primary removal mechanisms for STH. Filtration and adsorption by biological films on the substrates and plant roots in subsurface flow CW occur by attachment of helminth eggs to the substrates, plant roots, or substrate-plant roots complex. Sedimentation, filtration, and adsorption do not involve either inactivation or destruction of the eggs, but they separate the eggs from wastewater. The separated eggs remain in the sludge of WSP and free water surface flow CW or attached to biofilms on substrate and plant roots of subsurface flow CW allowing the effluents to be free of helminth eggs. Other removal mechanisms that apply in both WSP and CW systems include natural die off, predation, and chemicals such as ammonia. However, these mechanisms have little contribution [64].

Water turbulence, the number of ponds in a series, hydraulic retention time, sludge accumulation, and hydraulic short-circuiting are the factors affecting helminth removal in WSP systems. These factors affect the rate of helminth egg sedimentation. Water turbulence and overturning caused by water flow, wind, rain, human disturbance, buoyed gas babbles from pond sludge or temperature interfere with the gravitational settling of helminth eggs [4]. Long hydraulic retention time of wastewater in the system provides time for helminth egg sedimentation, while excessive accumulation of sludge affects pond hydraulics, creating short-circuiting that may carry helminth eggs through to the outlet or re-suspend eggs that have been deposited in the pond sediments. Increasing the number of ponds in a series increases helminth egg removal efficiency [65].

Hydraulic retention time and hydraulic short-circuiting also effect helminth egg removal in CW systems. As in WSP, long hydraulic retention times provide more time for helminth eggs to be exposed to the removal mechanisms such as sedimentation in free water flow CW or filtration in subsurface flow CW systems. In CW systems, hydraulic retention time depends on wastewater flow rate, water depth, vegetation, and type of substrate used. Hydraulic short-circuiting as a result of clogging at the inlet or outlet of a CW system may reduce wastewater residence time, therefore lowering helminth egg removal efficiency. Other factors affecting helminth egg removal in CW systems include the design or type of CW (subsurface systems have higher efficiency than surface systems) and vegetation coverage [66].

Country	System	STH species	Mean/MR influent (eggs/L)	Mean/MR effluent (eggs/L)	Reference
Burkina Faso	WSP	*A. lumbricoides*	7	0	[23]
		Hookworm	6	0	
		T. trichiura	1.4	0	
Egypt	CWs	*A. lumbricoides*	1.59	0	[49]
		Hookworm	0.12	0	
		T. trichiura	0.09	0	
		All STH	2.3	0	[41]
	WSP	*A. lumbricoides*	4	0	[50]
Kenya	WSPs	*A. lumbricoides*	17.5–133.5	0	[51]
		All STH	158–398	0	[52]
Morocco	WSPs	*A. lumbricoides*	0.4	0.01	[53]
		A. lumbricoides	4	0	[54]
		T. trichiura	2.2	0	
Nigeria	WSP	*A. lumbricoides*	12.38	0.19	[29]
		A. lumbricoides	7.69	0.19	
		T. trichiura	4.12	0.31	
Tanzania	WSPs	*A. lumbricoides*	10–19	0	[31]
		Hookworm	9.5–32	0.2–7.5	
		T. trichiura	20	0	
Tunisia	WSPs	*A. lumbricoides*	413.5–731	0–111.5	[46]
	CWs	*A. lumbricoides T. trichiura and E. vermicularis*	3.8	0.1–0.8	[55, 56]
Iran	CWs	*A. lumbricoides*	30.43	0.08	[57, 58]
	WSPs	*A. lumbricoides*	30–38	0	[58]
		T. trichiura	2.5	0	
Brazil	WSPs	All STH	992.6–1740	0	[41, 59]
Bolivia	WSP	*A. lumbricoides*	306	0	[60]
Cayman Island	WSP	*A. lumbricoides*	32	0	[47]
		Hookworm	113–957	33–690	
		T. trichiura	273	0	
Colombia	WSP	*A. lumbricoides*	183	0	[61]
		T. trichiura	31	0	
Honduras	WSPs	All STH	9–744	0	[62]
Spain	WSP	*Ascaris* spp. and *T. trichiura*	1.8	0	[63]

STH—Soil-transmitted helminths, MR—Range of means reported from different treatment systems in a particular country, WSP—Waste stabilization pond, and CW—Constructed wetland.

Table 2.
Concentration of soil-transmitted helminth eggs in the influents and effluents of waste stabilization ponds and constructed wetlands systems in different countries.

Due to their cost-effectiveness, natural wastewater treatment systems are preferred wastewater treatment systems in many low- and middle income-countries. Adequate maintenance and operation are critical to the performance of natural wastewater treatment systems. However, all too often these systems become overloaded and receive inadequate maintenance. Most factors associated with poor performance of natural wastewater treatment systems are the result of lack of adequate maintenance and repair, abandonment of the systems, or poor design [67]. Inadequate maintenance such as desludging results in sludge accumulation in the systems, which will reduce hydraulic residence of wastewater and sometimes create hydraulic short-circuiting resulting in poor performance of the systems. In CW systems, accumulation of sludge may result in clogging of the system leading to the system malfunctioning.

Generally, natural wastewater treatment systems receive influent wastewater with high concentrations of STH eggs and are capable of producing effluents containing $\leq$0.01 egg per liter, which is suitable for use or discharge to the environment even when children aged less than 15 years are exposed. The main reason for inadequate maintenance of natural wastewater treatment systems in low-income countries is a decrease in governmental financial support as well as decrease in finance generated by the systems as they become older [37]. In addition to that, poor system design such as errors in system geometry (e.g., length-width ratio) or poor arrangements of inlet and outlet may lead to water turbulence and hydraulic short-circuiting resulting in low system performance [60, 67].

5. Conclusion

Countries implementing prophylactic chemotherapy for controlling helminthiasis report high concentrations of STH eggs in wastewater. For the wastewater to be safe for reuse and/or discharge, it requires further treatment. Natural wastewater treatment systems including sedimentation ponds and constructed wetlands work well in assisting STH control through interrupting transmission by removing eggs from wastewater.

Conflict of interest

The authors declare that there are no conflicts of interests.

Author details

Abdallah Zacharia[1]*, Anne H. Outwater[2] and Rob Van Deun[3]

1 Department of Parasitology and Medical Entomology, Muhimbili University of Health and Allied Sciences, Dar es Salaam, Tanzania

2 Department of Community Health Nursing, Muhimbili University of Health and Allied Sciences, Dar es Salaam, Tanzania

3 Unit Life Sciences and Chemistry, Thomas More University of Applied Sciences, Geel, Belgium

*Address all correspondence to: naayz@ymail.com

References

[1] Zhang Y, Shen Y. Wastewater irrigation: Past, present, and future. Wiley Interdisciplinary Reviews Water. 2017:e1234. DOI: 10.1002/wat2.1234

[2] Chen Y, Wen Y, Tang Z, Li L, Cai Y, Zhou Q. Removal processes of disinfection byproducts in subsurface-flow constructed wetlands treating secondary effluent. Water Research. 2014;**51**:163-171. DOI: 10.1016/j.watres.2013.12.027

[3] Salgot M, Folch M. Wastewater treatment and water reuse. Current Opinion in Environmental Science and Health. 2018;**2**:74. DOI: 10.1016/j.coesh.2018.03.005

[4] Sengupta ME, Thamsborg SM, Andersen TJ, Olsen A, Dalsgaard A. Sedimentation of helminth eggs in water. Water Research. 2011;**45**:4651-4660. DOI: 10.1016/j.watres.2011.06.017

[5] IWA. The Wastewater Report 2017: Reuse Opportunity. London, United Kingdom: The Intragenerational Water Association; 2018

[6] Zurita F, White JR. Comparative study of three two-stage hybrid ecological wastewater treatment systems for producing high nutrient, reclaimed water for irrigation reuse in developing countries. Water. 2014;**6**:213-228. DOI: 10.3390/w6020213

[7] Okoh AI, Odjadjare EE, Igbinosa EO, Osode AN. Wastewater treatment plants as a source of microbial pathogens in receiving watersheds. African Journal of Biotechnology. 2007;**6**:2932-2944

[8] Stott R. Fate and behaviour of parasites in wastewater treatment systems. In: Mara D, Horan N, editors. Handbook of Water and Wastewater Microbiology. London: Academic Press; 2003. pp. 491-521

[9] Zacharia A, Outwater AH, Ngasala B, Van Deun R. Pathogenic parasites in raw and treated wastewater in Africa: A review. Resources and Environment. 2018;**8**:232-240. DOI: 10.5923/j.re.20180805.03

[10] Salama Y, Chennaoui M, El Amraoui M, Mountadar M, Rihani M, Assobhei O. Evaluation of the qualitative and quantitative of parasitic load and environmental risks of raw sewage rejected on the coast of the City of El Jadida (Morocco). Journal of Materials and Environmental Science. 2016;**7**:1614-1623

[11] Hindiyeh MY, Al-Salem S. Integrated guide to sanitary parasitology. Amman. World Health Organization Regional Office for the Eastern Mediterranean Regional Centre for Environmental Health Activities; 2004

[12] Ensink JHJ, Van der Hoek W, Mukhtar M, Tahir Z, Amerasinghe FP. High risk of hookworm infection among wastewater farmers in Pakistan. Transactions of the Royal Society of Tropical Medicine and Hygiene. 2005;**99**:809-818

[13] Habbaria K, Tifnoutib A, Bittonc G, Mandil A. Geohelminthic infections associated with raw wastewater reuse for agricultural purposes in Beni-Mellal, Morocco. Parasitology International. 2000;**48**:249-254

[14] Amoah ID, Abubakari A, Stenström TA, Abaidoo RC, Seidu R. Contribution of wastewater irrigation to soil transmitted helminths infection among vegetable farmers in Kumasi, Ghana. PLOS Neglected Tropical Diseases. 2016;**10**:e0005161. DOI: 10.1371/journal.pntd.0005161

[15] Ziegelbaure K, Speich B, Mausezahl D, Bos R, Keiser J, Utzinger J. Effect of sanitation on

soil-transmitted helminth infection: Systematic review and meta-analysis. PLoS Medicine. 2012;**9**:1-7. DOI: 10.1371/journal.pmed.1001162

[16] UN Water. The United Nations World Water Development Report 2017: Facts and Figures. Colombella, Italy: The United Nations; 2017

[17] Pescod MB. Wastewater characteristics and effluent quality parameters. In: Wastewater Treatment and Use in Agriculture - FAO Irrigation and Drainage Paper 47. Rome: Food and Agriculture Organization of the United Nations; 1992

[18] Liu L, Hall G, Champagne P. Effects of environmental factors on the disinfection performance of a wastewater stabilization pond operated in a temperate climate. Water. 2016;**8**:11. DOI: 10.3390/w8010005

[19] Hasegawa M, Pilotte N, Kikuchi M, Means AR, Papaiakovou M, Gonzalez AM, et al. What does soil-transmitted helminth elimination look like? Results from a targeted molecular detection survey in Japan. Parasites & Vectors. 2020;**13**:1-11. DOI: 10.1186/s13071-019-3875-z

[20] Paniker CJ. Textbook of Medical Parasitology. 6th ed. New Delhi: Jaypee Brothers Medical Publishers (P) LTD; 2007

[21] World Health Organization. 2030 Targets for Soil-Transmitted Helminthiases Control Programmes. Geneva: World Health Organization; 2019

[22] Jimenez-Cisneros B, Maya-Rendon C. Helminths and sanitation. In: Mendez-Vilas A, editor. Communicating Current Research and Educational Topics and Trends in Applied Microbiology. Formatex; 2007. pp. 60-71

[23] Konaté Y, Maiga AH, Basset D, Casellas C, Picot B. Parasite removal by waste stabilisation pond in Burkina Faso, accumulation and inactivation in sludge. Ecological Engineering. 2013;**50**:101-106. DOI: 10.1016/j.ecoleng.2012.03.021

[24] Gideon A, Fotseu A, Arnold K. Dissemination of the resistant forms of intestinal worms in the marshy areas of the city of Yaounde (Cameroon): Importance of some abiotic factors of the medium. Applied Water Science. 2019;**9**:1-9. DOI: 10.1007/s13201-019-0895-y

[25] Nzouebet WAL, Noumsi IM, Rechenburg A. Prevalence and diversity of intestinal helminth eggs in pit latrine sludge of a tropical urban area. The Journal of Water, Sanitation & Hygiene. 2016;**6**:622-630. DOI: 10.2166/washdev.2016.074

[26] Amoah ID, Reddy P, Seidu R, Stenström TA. Removal of helminth eggs by centralized and decentralized wastewater treatment plants in South Africa and Lesotho: Health implications for direct and indirect exposure to the effluents. Environmental Science and Pollution Research. 2018;**25**:12883-12895. DOI: 10.1007/s11356-018-1503-7

[27] Kumwenda S, Msefula C, Kadewa W, Ngwira B, Morse T. Estimating the health risk associated with the use of ecological sanitation toilets in Malawi. Journal of Environmental and Public Health. 2017;**2017**:1-13

[28] Kumwenda S, Msefula C, Kadewa W, Diness Y, Kato C, Morse T, et al. Is there a difference in prevalence of helminths between households using ecological sanitation and those using traditional pit latrines? A latrine based cross sectional comparative study in Malawi. BMC Research Notes. 2017;**10**:1-9. DOI: 10.1186/s13104-017-2519-7

[29] Okojokwu OJ, Inabo HI, Yakubu SE. Parasitology of Wastewater: Evaluation

Using Biosand Filter and Waste Stabilisation Pond. Saarbrucken: Lambert Academic Publishing; 2012

[30] Amoah ID, Reddy P, Seidu R, Stenström TA. Concentration of soil-transmitted helminth eggs in sludge from South Africa and Senegal: A probabilistic estimation of infection risks associated with agricultural application. Journal of Environmental Management. 2018;**206**:1020-1027. DOI: 10.1016/j.jenvman.2017.12.003

[31] Zacharia A, Ahmada W, Outwater AH, Ngasala B, Van Deun R. Evaluation of occurrence, concentration, and removal of pathogenic parasites and fecal coliforms in three waste stabilization pond systems in Tanzania. Scientific World Journal. 2019;**2019**:1-12. DOI: 10.1155/2019/3415617

[32] Fuhrimann S, Stalder M, Winkler MS, Niwagaba CB, Babu M, Masaba G, et al. Microbial and chemical contamination of water, sediment and soil in the Nakivubo wetland area in Kampala, Uganda. Environmental Monitoring and Assessment. 2015;**187**: 1-15. DOI: 10.1007/s10661-015-4689-x

[33] Grego S, Barani V, Hegarty-craver M, Raj A, Perumal P, Berg AB, et al. Soil-transmitted helminth eggs assessment in wastewater in an urban area in India. Journal of Water and Health. 2018;**16**:34-43. DOI: 10.2166/wh.2017.147

[34] Muntalif BS, Firdayati M, Lesmono FD, Sally A, Siregar V, Astrini P. Helminth eggs assessment of fecal sludge in urban area of Bandung, Indonesia. E3S Web of Conferences. 2020;**2**:1-6

[35] Firdayati M, Notodarmojo PA, Muntalif BS, Trihartomo D, Fathuna IS, Somantri K. Helminth eggs parameter of water spinach agriculture field in Bandung. Indonesian Journal of Urban and Environmental Technology. 2018;**2**:27-34. DOI: 10.25105/urbanenvirotech.v2i1.3553

[36] Guadagnini RA, Urbano L, Maura R, Franco B, Guimarães JR. Inactivation of bacteria and helminth in wastewater treatment plant effluent using oxidation processes. Water Science and Technology. 2013;**68**:1825-1829. DOI: 10.2166/wst.2013.431

[37] Verbyla ME. Pathogen removal in natural wastewater treatment and resource recovery systems: Solutions for small cities in an urbanizing world [thesis]. University of South Florida; 2015

[38] Ortiz C, Lopezy MC, Rivas FA. Helminth prevalence in a waste-water plant at El Rosal, Cundinamarca. Revista de Salud Pública. 2012;**14**:296-304

[39] Yaya-Beas R-E, Cadillo-La-Torre E-A, Kujawa-Roeleveld K, van Lier JB, Zeeman G. Presence of helminth eggs in domestic wastewater and its removal at low temperature UASB reactors in Peruvian highlands. Water Research. 2016;**90**:286-293. DOI: 10.1016/j.watres.2015.11.023

[40] Mahmood Q, Pervez A, Zeb BS, Zaffar H, Yaqoob H, Waseem M, et al. Natural treatment systems as sustainable ecotechnologies for the developing countries. BioMed Research International. 2013;**19**:1-19

[41] Stott R, May E, Mara D. Parasite removal by natural wastewater treatment systems: Performance of waste stabilisation ponds and constructed wetlands. Water Science and Technology. 2003;**48**:97-104

[42] Van Deun R. Short Training Initiative "Wetlands in Environmental Management" - 1- Treatment Wetlands. Geel, Belgium: Thomas More University of Applied Sciences; 2015. pp. 1-199

[43] Vymazal J, Greenway M, Tonderski K, Brix H, Mander U.

Constructed wetlands for wastewater treatment. In: JTA V, Beltman B, Bobbink R, Whigham DF, editors. Ecological Studies. Vol. 190. Berlin Heidelberg: Springer; 2006. pp. 69-96

[44] World Health Organization. Guideline for the Safe Use of Wastewater, Excreta and Greywater - Policy and Regulatory Aspects. Vol. I. Geneva: WHO; 2006

[45] Jimenez B. Helminth ova removal from wastewater for agriculture and aquaculture reuse. Water Science and Technology. 2007;**1**:485-493. DOI: 10.2166/wst.2007.046

[46] Ben Ayed L, Schijven J, Alouini Z, Jemli M. Presence of parasitic protozoa and helminth in sewage and efficiency of sewage treatment in Tunisia. Parasitology Research. 2009;**105**:393-406. DOI: 10.1007/s00436-009-1396-y

[47] Ellis KV, Rodrigues PCC, Gomez CL. Parasite ova and cysts in waste stabilization ponds. Water Research. 1993;**27**:1455-1460

[48] OutwaterAH,ZachariaA,NyomoraA, Francis F, Katima JHY. Potential of adding constructed wetland/fishpond to waste stabilization ponds to improve Helminths removal in Morogoro municipality, Tanzania. Journal of Applied Sciences and Environmental Management. 2019;**23**:2201-2203. DOI: 10.4314/jasem.v23i12.18

[49] Stott R, Jenkins T, Shabana M, May E. A survey of the microbial quality of wastewaters in Ismailia, Egypt and the implications for wastewater reuse. Water Science and Technology. 1997;**35**:211-217. DOI: 10.1016/S0273-1223(97)00261-8

[50] Shereif MM, Easa ME-S, El-Samra MI, Mancy KH. A demonstration on wastewater treatment for reuse applications in fish production and irrigation in Suez, Egypt. Water Science and Technology. 1995;**32**:137-144. DOI: 10.1016/0273-1223(96)00127-8

[51] Grimason AM, Smith HV, Young G, Thitai WN. Occurrences and removal of Ascaris sp. ova by waste stabilization ponds in Kenya. Water Science and Technology. 1996;**33**:75-82. DOI: 10.1016/0273-1223(96)00341-1

[52] Ayres RM, Mara DD, Lee DL, Thitai WN. Monitoring full scale waste stabilisation ponds in Kenya for nematode egg removal. Environmental Technology. 1993;**14**:295-300. DOI: 10.1080/09593339309385293

[53] Hajjami K, Ennaji MM, Fouad S, Oubrim N, Khallayoune K, Cohen N. Assessment of Helminths health risk associated with reuse of raw and treated wastewater of the Settat City (Morocco). Resources and Environment. 2012;**2**:193-201. DOI: 10.5923/j.re.20120205.03

[54] Bouhoum K, Amahmid O, Asmama S. Occurrence and removal of protozoan cysts and helminth eggs in waste stabilization ponds in Marrakech. Water Science and Technology. 2000;**42**:159-164. DOI: 10.2166/wst.2000.0632

[55] Mandi L, Bouhoum K, Ouazzani N. Application of constructed wetland for domestic wastewater treatment in an arid climate. Water Science and Technology. 1998;**38**:379-387. DOI: 10.1016/S0273-1223(98)80004-8

[56] Mandi L, Houhoum B, Asmama S, Schwartzbrod J. Wastewater treatment by reed beds an experimental approach. Wet Resin. 1996;**30**:2009-2016

[57] Sharafi K, Moradi M, Khosravi T, Pirsaheb M. A survey of parasitic contamination in Qasr-e-Shirin raw wastewater: The role of a constructed wetland in removing contamination.

International Journal of Health and Life Sciences. 2015;**1**:29-34

[58] Sharafi K, Fazlzadehdavil M, Pirsaheb M, Derayat J, Hazrati S. The comparison of parasite eggs and protozoan cysts of urban raw wastewater and efficiency of various wastewater treatment systems to remove them. Ecological Engineering. 2012;**44**:244-248. DOI: 10.1016/j.ecoleng.2012.03.008

[59] Dixo NGH, Gambrill MP, Catunda PFC, Van Haandel AC, De Engenharia A, Departamento A, et al. Removal of pathogenic organisms from the effluent of an upflow anaerobic digester using waste stabilization ponds. Water Science and Technology. 1995;**31**:275-284. DOI: 10.1016/0273-1223(95)98379-B

[60] Verbyla ME, Oakley SM, Lizima LA, Zhang J, Iriarte M, Tejada-Martinez AE, et al. *Taenia* eggs in a stabilization pond system with poor hydraulics: Concern for human cysticercosis? Water Science and Technology. 2013;**68**:2698-2703. DOI: 10.2166/wst.2013.556

[61] Madera CA, Pena MR, Mara DD. Microbiological quality of a waste stabilization pond effluent used for restricted irrigation in Valle Del Cauca, Colombia. Water Science and Technology. 2002;**45**:139-143

[62] Oakley SM. Monitoring of Wastewater Stabilization Ponds in Honduras. San Juan: Congreso Interamericano de Ingenieria Sanitaria y Ambiental; 2004. pp. 22-27

[63] Reinoso R. Pathogen Removal Mechanisms in Wastewater Treatment Systems [thesis]. University of León; 2008

[64] Verbyla M, von Sperling M, Maiga Y. Waste stabilization ponds. In: Rose JB, Jimenez B, editors. Global Water Pathogen Project Part Four. Management of Risk from Excreta and Wastewater. E. Lansing, MI: Michigan State University, UNESCO; 2017

[65] Verbyla ME, Iriarte MM, Mercado Guzmán A, Coronado O, Almanza M, Mihelcic JR. Pathogens and fecal indicators in waste stabilization pond systems with direct reuse for irrigation: Fate and transport in water, soil and crops. Science of the Total Environment. 2016;**551-552**:429-437. DOI: 10.1016/j.scitotenv.2016.01.159

[66] Maiga Y, von Sperling M, Mihelcic J. Constructed wetlands. In: Rose JB, Jimenez B, editors. Global Water Pathogen Project Part Four. Management of Risk from Excreta and Wastewater. E. Lansing, MI: Michigan State University, UNESCO; 2017

[67] Peña Varón M, Mara D. Waste Stabilisation Ponds. IRC International Water and Sanitation Centre; 2004

Chapter 4

Association of Polyethylene Glycol Solubility with Emerging Membrane Technologies, Wastewater Treatment, and Desalination

Hongbo Du, Subhani Bandara, Laura E. Carson and Raghava R. Kommalapati

Abstract

Forward osmosis (FO) and membrane distillation (MD) are two emerging membrane technologies, and both have advantages of low membrane fouling, ability to use for highly saline desalination, and feasibility to integrate with a low-grade heat source like solar collector. Because polyethylene glycol (PEG) is a flexible, water-soluble polymer, it is an essential material used for membrane fabrication and enhancement of membrane properties. Low-molecular-weight PEG sometimes is used as pore constrictor and pore former for developing MD membranes and support layer of FO membranes. Due to the affinity of PEG chains to water molecules, PEG, its derivatives, and copolymers have been widely used in the fabrication/ modification of FO and MD membranes, which are currently applied to bioseparation, wastewater treatment, and desalination in academia and industry at the pilot scale. This chapter covers direct PEG and its membrane separation applications in wastewater treatment and desalination. The advancement of PEG in membrane science and engineering is reviewed and discussed comprehensively. We focus on the effectiveness of PEG on membrane antifouling and the stability of PEG-modified membranes when applied to wastewater treatment and desalination.

Keywords: polyethylene glycol, forward osmosis, membrane distillation, draw solute, pore-forming additive

1. Chemistry of polyethylene glycol, its derivatives, and copolymers

Poly(ethylene glycol) or PEG is a synthetic water-soluble polymer with the formula $C_{2n}H_{4n+2}O_{n+1}$ which is available in a wide range of molecular weights where n value can go up to thousands. The molecular weight of PEG has a significant effect on its properties. Low-molecular-weight compounds (molecular weight < 1000) exist in liquid form, whereas higher molecular weight compounds are in waxlike solid form. The highest melting point of the solid-state material is reached at around 67°C, depending on the molecular weight [1]. In certain instances, PEG is also denoted as poly(oxyethylene) (POE) and polyoxirane. Additionally, PEG is called poly(ethylene oxide) (PEO), when the molecular weight exceeds 20000. Other than water PEG is also soluble in certain organic solvents like acetonitrile, ethylene dichloride, carbon tetrachloride, trichloroethylene, methylene dichloride, and dimethylformamide [2]. The favorable properties of PEG such as solubility in both aqueous and organic solvents, nontoxicity, and reduced immunogenicity have made it a good candidate to be used for conjugation with other molecules [3].

PEG is available as linear or branched chain polymers with terminal hydroxyl groups. The structure of PEG allows the attachment of varying functional groups at the end groups of the polymer. The attachment of different molecules to PEG is known as PEGylation, and it provides means of improving the solubility, stability, and biocompatibility of the attached molecules/compounds. PEG is commonly synthesized starting with ethylene oxide via an anionic ring-opening reaction, through a nucleophilic attack on the epoxide ring by the hydroxide ion. Conjugation of PEG to other molecules can be categorized into two groups as (1) first-generation PEGylation and (2) second-generation PEGylation. The first-generation PEGylation involves random attachment of PEG polymers to other molecules and is widely used with modifying polypeptides. This method can usually generate various undesired products as the attachment is nonselective. Also, it is mostly limited to low-molecular-weight derivatives and unstable bonds. On the other hand, the second-generation PEGylation is site-specific and leads toward the production of more stable and pure derivatives [4].

1.1 Hetero- and homobifunctional PEG derivatives

Various hetero- and homobifunctional products of PEG can be synthesized by different methods. Bentley et al. have shown a method to synthesize heterobifunctional PEG derivatives in high purity and high yield, by going through an intermediate with an easily removable group [5]. Here, they have first attached a benzyloxy group as the removable group to one end of PEG. Then, after modifying the other terminal OH group with a required molecule, the first group was removed by hydrogenolysis or hydrolysis. Afterward, another functional group can be attached to the newly available OH group, or the new OH can be converted to a different functional group. Also, another group has synthesized a heterobifunctional PEG with acetal and thiol groups starting with polymerization of ethylene oxide with potassium 3,3-diethoxypropanolate. Then, an excess of methansulfonyl chloride was used to convert a terminal alkoxide group to a methansulfonyl groups [6]. Within the two procedures above, the polymerization-based process is the most frequently used method in the synthesis of heterobifunctional derivatives. Although the second method is more cost-effective and efficient than the intermediate based method, it requires the availability of proper anionic polymerization initiators and precautions to avoid the formation of PEG diols [4].

A homobifunctional PEG derivative of α-lipoic acid (LA) ester was synthesized by Lu et al. to improve its properties for potential medical applications [7]. In this synthesis an esterification reaction driven by 1-ethyl-3-(3-dimethylaminopropyl) carbodiimide was carried out with PEG in the presence of 4-(dimethylamino) pyridine as a catalyst. Additionally, homobifunctional PEG has been used in metal nanoparticle synthesis as a stabilizing agent. For instance, Ge et al. have synthesized supramagnetic nanoparticles to be used as draw solutes in forward osmosis (FO) membrane using polyethylene glycol activated with two carboxylic acid groups at the terminal ends [8].

1.2 Monofunctional PEG derivatives

Monomethoxy PEG (mPEG), where one terminal of the PEG is capped with a relatively inert methoxy group, is commonly utilized in producing monofunctional PEG derivatives. The synthesis of mPEG is carried out via anionic ring-opening polymerization reaction initiated by methoxide ions. The presence of trace amounts of water during the synthesis process of mPEG can result in the formation of PEG diols, which can reach above 15% in the composition. Hence, during the synthesis of monofunctional PEG, necessary steps must be taken to remove PEG diols from the starting materials. Otherwise, the final product will contain bifunctional PEG as impurities. Therefore, conversion of diols to inert compounds such as PEG-dimethyl ether or PEG carboxylic acids followed by purification was used as a strategy to overcome this issue [4, 9].

A group of monofunctional PEG derivatives called NHS esters, where N-hydroxysuccinimide based group is attached to mPEG, are widely used in

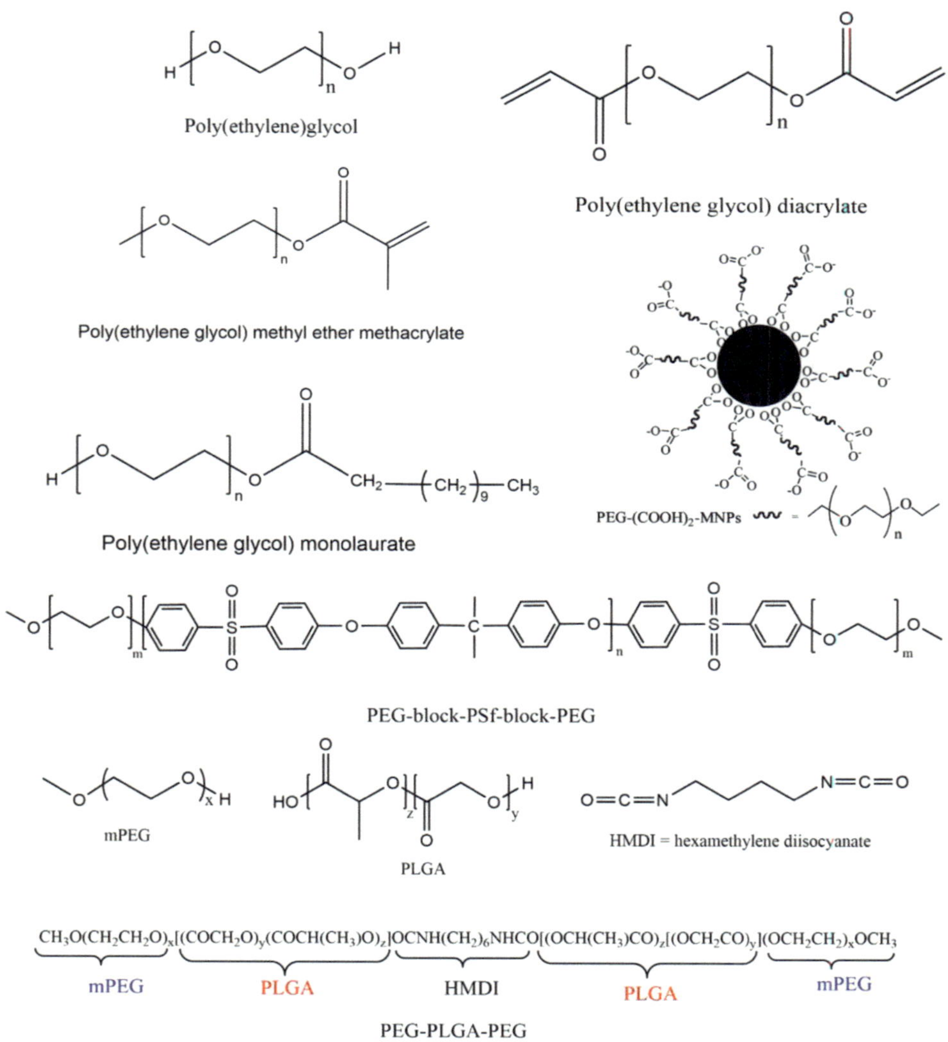

Figure 1.
Structures of PEG, PEG derivatives, and copolymers used in membranes (PEG-(COOH)-MNPs adapted from [8]).

protein and peptide modifications. This type of PEG derivatives is commonly used as acylating agents to modify amino groups of lysine residues and also has some reactivity with histidine imidazole and tyrosine hydroxyl groups. For example, PEG succinimidyl succinate (PEG-SS) produced by reacting mPEG with succinic anhydride followed by carboxylic acid activation to form succinimidyl ester is an NHS ester that has been successfully coupled to the enzyme asparaginase [4, 9]. Also, PEG derivatives such as trichlorophenyl carbonate and carbonylimidazole were synthesized by reacting mPEG hydroxyl group with chloroformates or carbonylimidazole. Alkylating reagents derived from mPEG include PEG tresylate and PEG dichlorotriazine. Other examples of monofunctional PEG derivatives that are specific to sulfhydryl groups include PEG-maleimide, PEG-vinylsulfone, and PEG-iodoacetamide [10]. In addition to protein modifications, monofunctional PEG derivatives were also used in osmosis membrane-related applications. For instance, PEG conjugated to fatty acid and PEG monolaurate was used as draw solutes to test forward osmosis membranes [10].

According to the abovementioned chemistry and properties of PEG, it has been widely used in many different areas such as biomedical, biotechnology, and membrane technology-based applications. The main focus of this chapter is to discuss the usage of PEG, its derivatives, and copolymers (**Figure 1**) in emerging membrane technologies, such as forward osmosis and membrane distillation, as their applications relate to wastewater treatment and desalination.

2. Principle of forward osmosis

Forward osmosis is an emerging technology using a semipermeable ultrathin membrane to treat water or wastewater, and the membrane is typically thinner than RO membranes. Similar to the structure of thin film RO membranes, an FO membrane typically consists of an active layer and a support layer. In a commercially available thin film composite (TFC) FO membrane, the active layer is polyamide (PA), and the support layer is mainly made of polysulfone (PSf) or polyethersulfone (PES). An FO system generally consists of an FO module/cell for holding FO membranes, a draw solute recovery unit and pumps for circulating feed and draw solution (**Figure 2**). The FO module/cell is classified into flat sheet, hollow fiber, and spiral wound configurations, mainly depending on the operating scale. Although FO takes some advantages of an osmotically driven process, such as less membrane fouling, low energy requirement and operating cost, over

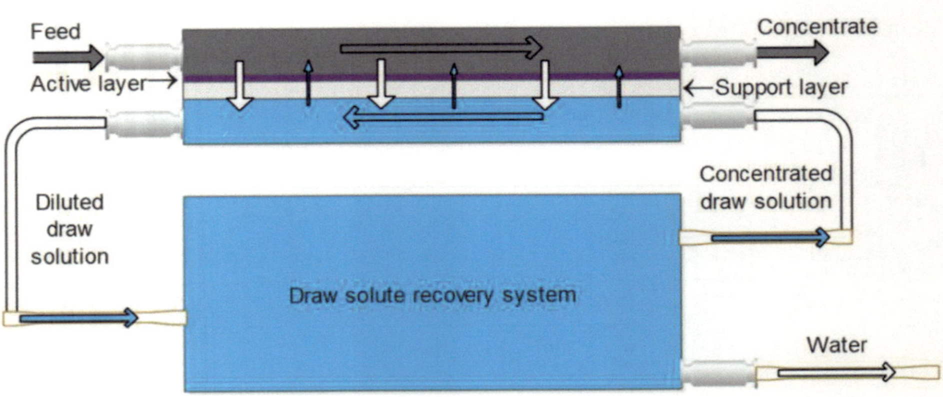

Figure 2.
Schematic of a forward osmosis system with draw solute recovery.

pressure-driven processes of nanofiltration (NF), and reverse osmosis (RO) in a similar membrane separation range, it still encounters some challenges in a wide range of practical applications for wastewater treatment and desalination. Jampani and Raghavarao purified and concentrated red cabbage (*Brassica oleracea* L.) anthocyanins by integrating osmotic membrane distillation and FO processes with a PEG-4000 and $MgSO_4$ (14.8/10.3%, w/w) extraction system [11]. Later, they used similar technologies to separate and concentrate Jamun anthocyanins by using PEG 6000 in the extraction system [12]. To test the performance of FO membranes, a series of PEG (10 and 20 kDa) and PEO (100, 600 and 1000 kDa) were used to determine molecular weight cutoff (MWCO), pore size, and distribution of a hydrophilic support layer during the fabrication of a TFC FO membrane [13].

3. Principle of membrane distillation

Membrane distillation (MD) is another emerging technology in membrane separation, which uses a hydrophobic, microporous membrane to treat water or wastewater through a thermally driven separation process. In an MD process, the

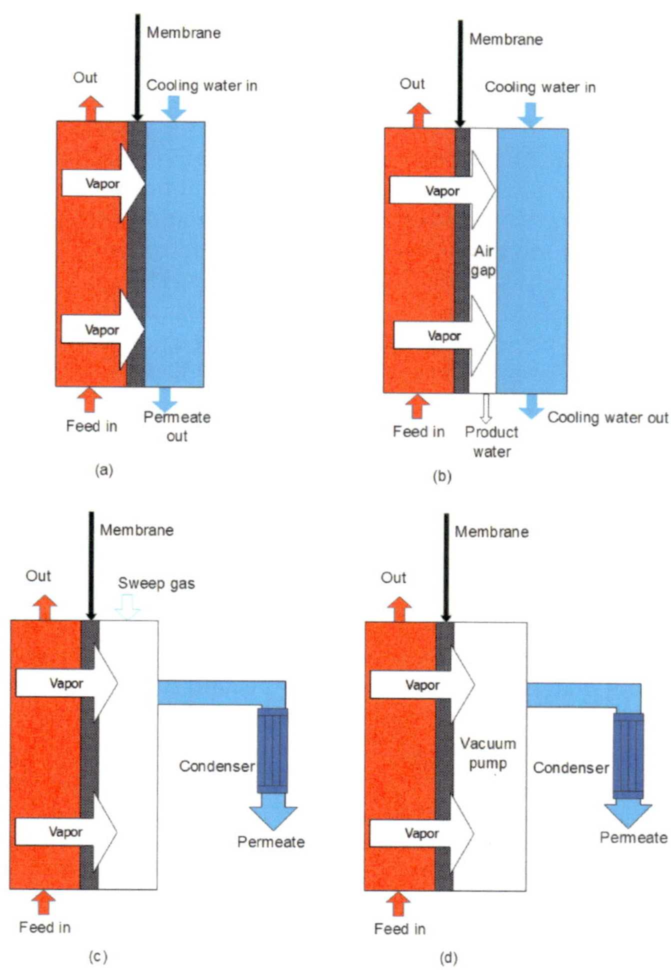

Figure 3.
Configurations of membrane distillation. (a) DCMD. (b) AGMD. (c) SGMD. (d) VMD.

feed solution is usually heated up, and the produced vapor passes through the pores of the membrane and condenses on the distillate side by cooling. MD is considered cost-effective and promising because it can achieve almost 100% dissolved solid rejection specifically in desalination. Similar to an FO module, an MD module can also be constructed as flat sheet, hollow fiber, and spiral wound forms. According to different configurations of the distillate side [14], MD can be classified into direct-contact MD (DCMD), air-gap MD, sweeping-gas MD (SGMD), and vacuum MD (VMD) shown in **Figure 3**. Conventional microporous membrane with pore size (0.1–1 μm) can be used for MD, and there also are some membranes specifically designed for MD. PEG and its derivatives can be used to improve the hydrophilicity of the membrane surface facing the feed solution during the fabrication of some specific MD membranes [15, 16].

4. PEG associated with forward osmosis

4.1 PEG used as/in draw solute

One of the challenges of FO applications in wastewater treatment and desalination is the selection of high-performance draw solutes. Among several hundreds of draw solute explored in the FO process, PEG is also evaluated for the FO process. Beside electrolyte NaCl, neutral PEG polymers at different molecular weights (M.W. = 100, 200, 600, 2000, 3000, 8300, and 10000) were used as model draw solute to evaluate the performance of a commercially available cellulose triacetate (CTA) FO membrane and a homemade porous UF-like FO membrane [17, 18]. Wei et al. fabricated a double-skinned selective thin film composite (TFC) FO membrane consisting of a top thin polyamide (PA) layer, a middle porous cellulose ester layer, and another bottom thin PA layer and tested its performance using several viscous draw solutes such as PEG monolaurate (PEG 640ML), sucrose, and ferric citric acid complex (Fe-CA) [19]. The novel membrane can minimize the effects of internal concentration polarization (ICP) because the bottom thin PA layer prevents viscous draw solute from entering the pores of the middle layer.

Hydrophilic magnetic nanoparticles (HMNPs) are a type of promising draw solutes, which may easily be recycled under a magnetic field. There exist some reports that HMNPs were fabricated from PEG and magnetic nanoparticles. Ge et al. synthesized a series of PEG-$(COOH)_2$-coated MNPs with narrow size distribution through a thermal decomposition process [8]. Mishra et al. specifically synthesized HMNPs with PEG 400 and evaluated their performances in an FO process where synthetic saline water (NaCl solution) at different concentrations of 0, 5, 10, 20 and 35 g/L was used as feed solution. About 35 g/L is close to the level of total dissolved solids (TDS) in seawater. When these HMNPs were used as draw solute in a fundamental FO process of deionized water, they could significantly eliminate the draw solute reverse diffusion problems which are common in the applications of general salts, such as NaCl, KCl, $MgCl_2$, $MgSO_4$, etc., as draw solute [20, 21]. Biodegradable and biocompatible temperature-sensitive triblock copolymer hydrogels PEG-PLGA-PEG/GO-0.09 wt%, PEG-PLGA-PEG/GO-0.18 wt%, PEG-PLGA-PEG/G-0.09 wt%, and PEG-PLGA-PEG/G-0.18 wt% were fabricated and used as draw solute in FO by Nakka and Mungray [22], where GO represents graphene oxide and G is graphene. PEG-PLGA-PEG was synthesized from D,L-lactide, 1,4-dioxane-2,5-dione, methyl ether polyethylene through ring-opening polymerization using stannous octane as catalyst. However, much smaller water fluxes were achieved when feed solutions are DI water and 2 g/L NaCl solutions than the previous HMNPs as draw solute.

4.2 PEG in the support layer of an FO membrane

In order to improve the performance of an FO membrane, the support layer can be reconstructed, and the active layer can be modified with PEG or its copolymer. Addition of PEG 400 to the support layer was conducted to fabricate a TFC FO membrane, and it was found that the addition of 6 wt% PEG was needed to reach the highest water flux [23] when DI water and 2 M NaCl were used as feed and draw solution. PEG 400 and dimethyl sulfone ($DMSO_2$) were used as an additive and a crystallizable diluent to fabricate the CTA support layer of an FO membrane through thermally induced phase separation, and the FO membrane exhibited better antifouling properties than PSf-based FO membranes [24]. Sharma et al. used PEG 4000 and 6000 as additive to prepare cellulose acetate flat asymmetric FO membranes, and the modified FO membranes were used to evaluate power density performance in pressure-retarded osmosis [25].

Liu et al. fabricated the support layer from PSf with 5-, 10-, and 15-wt% PEG or PEGMA (poly(ethylene glycol) methyl ether methacrylate) and evaluated the corresponding FO membranes by using DI water and 1 M NaCl as feed and draw solution [26]. The PSf-PEG support layer was made by blending PEG with PSf, and in the second type, it was PEGMA grafted on PSf. The FO membrane containing 10 wt% PEG achieved relatively steady performance for a long time operation process due to its better salt rejection, and the FO membrane with 5% PEGMA grafting possessed a high intrinsic permeability and a low structural parameter. Recently, amphiphilic PEG-block-PSf-block-PEG copolymers were used to cast the support layer, and the fabricated TFC FO membrane achieved some significant improvements on water flux, antifouling, and permeability selectivity [27].

4.3 PEG in the active layer of an FO membrane

When the active layer of a TFC FO membrane is modified with PEG, the surface hydrophilicity of the membrane can be improved, thus enhancing the membrane antifouling properties. Elimelech et al. functionalized the active layer of a TFC FO membrane with PEG diepoxides through surface grafting, and their dynamic experiments showed that the membrane fouling was significantly reduced when testing with alginate as model organic foulant [28]. The same research group later used a post-fabrication technique to graft a PEG-block copolymer on the active layer of commercial TFC FO membranes, and the PEG density was optimized to the best membrane performance by compromising the increased membrane hydrophilicity and reduced water flux [29]. Interestingly, a novel design of FO membranes by impregnating the support layer with hydrophilic cross-linked poly(ethylene glycol) diacrylate (PEGDA) was proposed by Zhao et al., and there is no additional PA layer needed [30]. The newly designed FO membrane had the ability to mitigate internal concentration polarization which is commonly for typical two-layer FO membranes and to improve the performance ratio by 50% compared to those of state-of-the-art commercial FO membranes. Recently, Chen et al. tethered the active PA layer of a TFC FO membrane with PEGMA, and the membrane ICP was greatly mitigated with only slightly flux reduction from 10.99 to 9.32 LMH during synthetic sewage treatment [31].

The antifouling ability and performance of CTA FO membranes can also be improved by applying PEG to the membrane surface. The surface of a CTA FO membrane was modified by firstly coating polydopamine (PD) and then grafting PEG, and the submerged osmotic membrane bioreactor using the FO membrane possessed better flux behaviors than the pristine reactor and anti-adhesion abilities of biopolymers and bio-cake [32].

5. PEG associated with membrane distillation

5.1 PEG used as pore-forming additive

In the early 1990s, PEG was used as pore-forming additive to fabricate microporous polyvinylidene fluoride (PVDF) membrane for MD of wastewater discharged from the taurine production [33]. The effects of a series of PEG 400, 1000, 1540, 2000, and 6000 on the pore structure and permeate performance of poly(vinylidene fluoride-co-hexafluoropropylene) (F2.6) flat-sheet membranes were investigated by evaluating average pore radius, porosity, and morphology, and the membranes reached better higher distilled flux than the PVDF membrane during the DCMD test [34]. Dayanandan et al. studied the influence of the various additions of PEG (0–4 wt%) in the coagulation bath composition during the preparation of PVDF membranes, and they found that the bath-based MD membrane with 4 wt% PEG had relative superior overall performance than other membranes based on various evaluations of elongation-at-break, tensile strength, liquid entry pressure, hydrophobicity, porosity, and water flux [35]. Combined effects of poly(vinylidene fluoride-co-hexafluoropropylene) (PVDF-HFP) and the concentrations of the additive PEG 10000 were studied by using a statistical approach, and the optimized membrane achieved salt rejection of 99.5% in the DCMD of 0.1 M NaCl solution at 65°C [36]. Two pore-forming additives PEG and LiCl both at 4 wt% were used in the fabrication of hydrophobic flat sheet and hollow fiber PVDF and PVDF-co-chlorotrifluoroethylene (PVDF-CTFE) membranes for membrane distillation [37, 38]. The effect of mass ratio of PEG and LiCl was further investigated for making PVDF-CTFE membranes for MD [39]. Recently, more effective additives such as organic acids, LiCl, $MgCl_2$, and $LiCl/H_2O$ mixtures along with PEG were investigated in the fabrication of flat-sheet hydrophobic PVDF-CTFE membranes used for MD [38]. Hou et al. prepared hydrophobic PVDF flat-sheet membranes for DCMD by using various non-solvent additives such as acetone, phosphoric acid, glycerine, LiCl, and PEG 400 [40]. They found that the membrane fabricated with 5 wt% acetone and 3 wt% phosphoric acid exhibited the highest water flux among the various fabricated membranes and showed great performance stability in the 240 h desalination of synthetic seawater. There exist more novel additives, such as calcium carbonate nanoparticles and TamiSolve® NxG along with PEG and LiCl, applied to the fabrication of hydrophobic membranes for MD [41, 42].

During the fabrication of microporous PVDF hollow fiber membranes, two non-solvent additives PEG 400/1500 and LiCl were added to the feed N,N-dimethylacetamide (DMAc) solution containing PVDF powder [43–46]. The experiment factors such as dope extrusion rate, take-up speed, air-gap concentration of polymer and additives, and bore liquid temperature were investigated in the membrane fabrication through a dry-jet wet phase inversion process, and the obtained membranes were suitable for DCMD than VMD. Recently, Zhang et al. further discovered the regulatory role of coagulation bath temperature during PVDF membrane fabrication when PEG 400 and triethyl phosphate were used [47]. In the fabrication process of PVDF hollow fiber membranes via complex thermally integrated phase separation at 80°C for MD, triethylphosphate (TEP) and PEG 200 were used as weak solvent and weak bore fluid, respectively [48]. PEG 6000 was used to fabricate hollow fiber MD membranes by Garcia-Payo et al. [49, 50]. They also dissolved PEG 10000 in DMAc to prepare PVDF-HFP hollow fiber membranes, and the optimized membrane had high permeate flux and salt rejection [51, 52]. Similarly, the concentration effects of PEG 600 (0–20 wt%) on the performance of PVDF-HFP membranes for MD were also investigated, and the suitable range of 5–20 wt% was found for fabricating hollow fiber MD membranes with acceptable performance [53]. Recently, Wang et al. developed hydrophobic flat sheet and hollow fiber membranes PVDF and PVDF-CTFE for DCMD using PEG 400 and LiCl as the additives [54].

5.2 PEG applied to surface modification of MD membrane

In order to explore new applications of microfiltration (MF) membrane in MD, a hydrophobic PVDF membrane with pore size of 0.2 μm and thickness of 125 μm was surface-casted with hydrophilic polymer gel made from a polymer solution containing 20% PEG and 3% polyvinyl alcohol (PVA) [15]. The modified membrane exhibited the excellent abilities to keep high constant flux at high salt concentration of 20–25% NaCl for a long time MD running and to prohibit the wetting problem when testing with 25% ethanol even. Recently, Zuo and Wang modified a PVDF membrane with a pore size of 0.22 μm and thickness of 125 μm by grafting PEG and depositing TiO_2 micro-balloon-shaped particles on the membrane surface [16]. In the desalination test of synthetic seawater containing 0.01 wt% of mineral oil over the 1-day operation, the fabricated membrane with a highly hydrophilic surface kept a stable water flux with negligible fouling and wetting. Later, Meng et al. investigated the effect of templating agents, such as PEG 1000, Pluronics F-127, Wacker IM-22, and cetyltrimethylammonium bromide, on the properties and MD performance of TiO_2-coated PVDF membranes [55]. Their MD membrane templated with PEG achieved the most promising overall performance of water flux, salt rejection, and extended operation time due to the optimum reduction of pore wetting induced by the templating agent PEG. They further developed superhydrophobic nanocomposite PVDF membranes for DCMD by modifying the conventional PVDF membrane with a pore size of 0.45 μm, and a dip coating of sol-gel containing PEG was used after a TiO_2 and fluro-silane coating in the modification [56].

6. PEG-assisted membranes in wastewater treatment and desalination

As emerging membrane technologies, FO and MD have received increased attentions for wastewater treatment and desalination [14, 57–60]. FO takes some

Membrane type	PEG	Wastewater/desalination	Performance improvement	Ref.
FO	PEGDE	Synthetic wastewater	Alleviation of flux reduction by 50%	[28]
FO	Jeffamine ED-2003	Synthetic secondary wastewater	Complete removal of organic foulant with NaCl solution	[29]
MD	PEG 2000	Taurine wastewater	Nonvolatile solute can be concentrated by MD	[33]
MD	PEG 2000 and 10000	NaCI solution	Remain 91% flux for 3.5 NaCl; almost unchanged flux for 20% NaCl	[15]
MD	PEG 1500	3.5% NaCl	Flux: 40.5 kg/m^2 h; NaCl rejection: 99.99% at 81.8°C	[45]
MD	PEG 1000	10% NaCl	Recovery ratio: about 68%	[55]
MD	PEG 200 and 400	Formulated seawater	Flux: 61.6 kg/m^2 h; NaCl rejection: 99.99% at 71°C	[48]

Table 1.
PEG-assisted FO and MD membranes used in wastewater treatment and desalination.

advantages of low membrane fouling, feasibility to treat high salinity water, possibility to run at low voltage electricity, and applications in osmotic dilution at low energy demand. When thermal volatile draw solute is used, FO can easily be integrated with low-grade heat, such as waste heat in a power plant and thermal heat gathered by highly efficient solar collector, for draw solute recovery [57, 58]. MD as a thermal membrane separation process has the nature to utilize low-grade heat easily, and it also possesses distinctive advantages of low membrane fouling and low operation energy demand of heating the feed and cooling the permeate when integrating with low-grade heat source and high quality of product water when using VMD and AGMD in the desalination of highly saline water [59]. The utilization of PEG, its derivatives, and copolymers in FO and MD membranes can improve the overall membrane performance, thus enhancing the abilities of these membranes in various applications. **Table 1** summarizes the performance of PEG-assisted FO and MD membranes used in wastewater treatment and desalination.

7. Summary

PEG, its derivatives, and copolymers are water soluble depending on their molecular weights, terminal, and copolymer blocks. They have widely been utilized in membrane fabrications and performance tests of some membranes due to the nature of hydrophilicity of these polymers. This chapter focuses on PEG applications in two emerging membrane technologies FO and MD. With regard to FO, PEG can not only be integrated into some draw solutes for easy draw solute recovery but can also be applied to the support and active layers for improving membrane antifouling properties. On the other hand, PEG is mainly used as pore-forming additive in the MD membrane fabrication. Compared to the control membranes in the various studies which are summarized here, PEG-assisted FO and MD membranes exhibit better overall performance for wastewater treatment and desalination according to water flux, flux recovery after cleaning, and antifouling behaviors.

Acknowledgements

This work is supported by the National Science Foundation (NSF) through the Center for Energy and Environmental Sustainability (CEES) at Prairie View A&M University, CREST Award #1036593.

Conflict of interest

The authors declare no conflict of interest.

Author details

Hongbo Du[1], Subhani Bandara[2], Laura E. Carson[2] and Raghava R. Kommalapati[3]*

1 Center for Energy and Environmental Sustainability, Prairie View A&M University, Prairie View, Texas, USA

2 Cooperative Agricultural Research Center, Prairie View A&M University, Prairie View, Texas, USA

3 Department of Civil and Environmental Engineering, Center for Energy and Environmental Sustainability, Prairie View A&M University, Prairie View, Texas, USA

*Address all correspondence to: rrkommalapati@pvamu.edu

References

[1] Harris JM. Poly(Ethylene Glycol) Chemistry: Biotechnical and Biomedical Applications. New York, USA: Springer Science & Business Media; 2013

[2] Bailey FE, Koleske JV. Poly(Ethylene Oxide). New York, USA: Academic Press; 1976

[3] Harris JM, Zalipsky S, Staff ACS. Poly(Ethylene Glycol). Vol. 680. Washington, DC: American Chemical Society; 1997

[4] Roberts MJ, Bentley MD, Harris JM. Chemistry for peptide and protein PEGylation. Advanced Drug Delivery Reviews. 2012;**64**:116-127. DOI: 10.1016/j.addr.2012.09.025

[5] Bentley MD, Harris JM, Kozlowski A. Heterobifunctional poly(ethylene glycol) derivatives and methods for their preparation. Google Patents. 2005

[6] Akiyama Y, Otsuka H, Nagasaki Y, Kato M, Kataoka K. Selective synthesis of heterobifunctional poly(ethylene glycol) derivatives containing both mercapto and acetal terminals. Bioconjugate Chemistry. 2000;**11**(6):947-950. DOI: 10.1021/bc000034w

[7] Lu C, Kim BM, Chai KY. Design, synthesis and evaluation of PEGylated lipoic acid derivatives with functionality as potent anti-melanogenic agents. European Journal of Medicinal Chemistry. 2011;**46**(10):5184-5188. DOI: 10.1016/j.ejmech.2011.07.056

[8] Ge QC, Su JC, Chung TS, Amy G. Hydrophilic superparamagnetic nanoparticles: Synthesis, characterization, and performance in forward osmosis processes. Industrial & Engineering Chemistry Research. 2011;**50**(1):382-388. DOI: 10.1021/ie101013w

[9] Turecek PL, Bossard MJ, Schoetens F, Ivens IA. PEGylation of biopharmaceuticals: A review of chemistry and nonclinical safety information of approved drugs. Journal of Pharmaceutical Sciences. 2016;**105**(2):460-475. DOI: 10.1016/j.xphs.2015.11.015

[10] Veronese FM. PEGylated Protein Drugs: Basic Science and Clinical Applications. Basel, Switzerland: Birkhäuser Basel, Springer; 2009

[11] Jampani C, Raghavarao K. Process integration for purification and concentration of red cabbage (*Brassica oleracea* L.) anthocyanins. Separation and Purification Technology. 2015;**141**:10-16. DOI: 10.1016/j.seppur.2014.11.024

[12] Jampani C, Raghavarao K. Separation and concentration of anthocyanins from jamun: An integrated process. Chemical Engineering Communications. 2015;**202**(10):1368-1379. DOI: 10.1080/00986445.2014.935351

[13] Wang YQ, Xu TW. Anchoring hydrophilic polymer in substrate: An easy approach for improving the performance of TFC FO membrane. Journal of Membrane Science. 2015;**476**:330-339. DOI: 10.1016/j.memsci.2014.11.025

[14] Gonzalez D, Amigo J, Suarez F. Membrane distillation: Perspectives for sustainable and improved desalination. Renewable & Sustainable Energy Reviews. 2017;**80**:238-259. DOI: 10.1016/j.rser.2017.05.078

[15] Peng P, Fane AG, Li XD. Desalination by membrane distillation adopting a hydrophilic membrane. Desalination. 2005;**173**(1):45-54. DOI: 10.1016/j.desal.2004.06.208

[16] Zuo GZ, Wang R. Novel membrane surface modification to enhance

anti-oil fouling property for membrane distillation application. Journal of Membrane Science. 2013;**447**:26-35. DOI: 10.1016/j.memsci.2013.06.053

[17] Yasukawa M, Tanaka Y, Takahashi T, Shibuya M, Mishima S, Matsuyama H. Effect of molecular weight of draw solute on water permeation in forward osmosis process. Industrial & Engineering Chemistry Research. 2015;**54**(33):8239-8246. DOI: 10.1021/acs.iecr.5b01960

[18] Qi SR, Li Y, Wang R, Tang CYY. Towards improved separation performance using porous FO membranes: The critical roles of membrane separation properties and draw solution. Journal of Membrane Science. 2016;**498**:67-74. DOI: 10.1016/j.memsci.2015.10.003

[19] Wei R, Zhang S, Cui Y, Ong RC, Chung TS, Helmer BJ, et al. Highly permeable forward osmosis (FO) membranes for high osmotic pressure but viscous draw solutes. Journal of Membrane Science. 2015;**496**:132-141. DOI: 10.1016/j.memsci.2015.08.054

[20] McCutcheon JR, Elimelech M. Influence of membrane support layer hydrophobicity on water flux in osmotically driven membrane processes. Journal of Membrane Science. 2008;**318**(1-2):458-466. DOI: 10.1016/j.memsci.2008.03.021

[21] Yang Q, Wang KY, Chung TS. Dual-layer hollow fibers with enhanced flux as novel forward osmosis membranes for water production. Environmental Science & Technology. 2009;**43**(8):2800-2805. DOI: 10.1021/es803360t

[22] Nakka R, Mungray AA. Biodegradable and biocompatible temperature sensitive triblock copolymer hydrogels as draw agents for forward osmosis. Separation and Purification Technology. 2016;**168**:83-92. DOI: 10.1016/j.seppur.2016.05.021

[23] Wu Y, Zhu HT, Feng L, Zhang LQ. Effects of polyethylene glycol on the structure and filtration performance of thin-film PA-Psf composite forward osmosis membranes. Separation Science and Technology. 2016;**51**(5):862-873. DOI: 10.1080/01496395.2015.1119846

[24] Yu Y, Wu QY, Liang HQ, Gu L, Xu ZK. Preparation and characterization of cellulose triacetate membranes via thermally induced phase separation. Journal of Applied Polymer Science. 2017;**134**(6). Article No: 44454. DOI: 10.1002/app.44454

[25] Sharma M, Mondal P, Chakraborty A, Kuttippurath J, Purkait M. Effect of different molecular weight polyethylene glycol on flat sheet cellulose acetate membranes for evaluating power density performance in pressure retarded osmosis study. Journal of Water Process Engineering. 2019;**30**. Article No: 4100632. DOI: 10.1016/j.jwpe.2018.05.011

[26] Liu BC, Chen C, Zhao PJ, Li T, Liu CH, Wang QY, et al. Thin-film composite forward osmosis membranes with substrate layer composed of polysulfone blended with PEG or polysulfone grafted PEG methyl ether methacrylate. Frontiers of Chemical Science and Engineering. 2016;**10**(4):562-574. DOI: 10.1007/s11705-016-1588-9

[27] An XC, Hu YX, Wang N, Wang T, Liu ZY. Breaking the permeability-selectivity trade-off in thin-film composite polyamide membranes with a PEG-b-PSF-b-PEG block copolymer ultrafiltration membrane support through post-annealing treatment. NPG Asia Materials. 2019;**11**. Article No: 13. DOI: 10.1038/s41427-019-0114-1

[28] Castrillon SRV, Lu XL, Shaffer DL, Elimelech M. Amine enrichment and poly(ethylene glycol) (PEG) surface modification of thin-film composite forward osmosis membranes for organic fouling control. Journal of Membrane

Science. 2014;**450**:331-339. DOI: 10.1016/j.memsci.2013.09.028

[29] Shaffer DL, Jaramillo H, Castrillon SRV, Lu XL, Elimelech M. Post-fabrication modification of forward osmosis membranes with a poly(ethylene glycol) block copolymer for improved organic fouling resistance. Journal of Membrane Science. 2015;**490**:209-219. DOI: 10.1016/j.memsci.2015.04.060

[30] Zhao SZ, Huang KP, Lin HQ. Impregnated membranes for water purification using forward osmosis. Industrial & Engineering Chemistry Research. 2015;**54**(49):12354-12366. DOI: 10.1021/acs.iecr.5b03241

[31] Chen Q, Sun F, Zhou J, Lu Y, Li YY, Yu HY, et al. Chlorine-resistant and internal-concentration-polarization-mitigated polyamide membrane via tethering poly(ethylene glycol) methacrylate. Journal of Applied Polymer Science. 2019;**136**(20). Article No: 47406. DOI: 10.1002/app.47406

[32] Li F, Cheng QX, Tian Q, Yang B, Chen QY. Biofouling behavior and performance of forward osmosis membranes with bioinspired surface modification in osmotic membrane bioreactor. Bioresource Technology. 2016;**211**:751-758. DOI: 10.1016/j.biortech.2016.03.169

[33] Wu Y, Kong Y, Liu J, Zhang J, Xu J. An experimental study on membrane distillation-crystallization for treating waste water in taurine production. Desalination. 1991;**80**(2-3):235-242. DOI: 10.1016/0011-9164(91)85160-V

[34] Feng CS, Wang R, Shi BL, Li GM, Wu YL. Factors affecting pore structure and performance of poly(vinylidene fluoride-co-hexafluoro propylene) asymmetric porous membrane. Journal of Membrane Science. 2006;**277**(1-2):55-64. DOI: 10.1016/j.memsci.2005.10.009

[35] Dayanandan N, Kapoor A, Karunanithi S, Sivaraman P. Studies on the influence of coagulation bath composition on the preparation of PVDF membranes. Desalination and Water Treatment. 2018;**122**:365-372. DOI: 10.5004/dwt.2018.23107

[36] Khayet M, Cojocaru C, Garcia-Payo MC. Experimental design and optimization of asymmetric flat-sheet membranes prepared for direct contact membrane distillation. Journal of Membrane Science. 2010;**351**(1-2): 234-245. DOI: 10.1016/j.memsci.2010.01.057

[37] Zheng LB, Wang J, Wei YS, Zhang Y, Li K, Wu ZJ. Interconnected PVDF-CTFE hydrophobic membranes for MD desalination: Effect of PEGs on phase inversion process. RSC Advances. 2016;**6**(25):20926-20937. DOI: 10.1039/c5ra28081h

[38] Zheng LB, Wu ZJ, Zhang Y, Wei YS, Wang J. Effect of non-solvent additives on the morphology, pore structure, and direct contact membrane distillation performance of PVDF-CTFE hydrophobic membranes. Journal of Environmental Sciences. 2016;**45**:28-39. DOI: 10.1016/j.jes.2015.09.023

[39] Zheng LB, Wu ZJ, Wei YS, Zhang Y, Yuan Y, Wang J. Preparation of PVDF-CTFE hydrophobic membranes for MD application: Effect of LiCl-based mixed additives. Journal of Membrane Science. 2016;**506**:71-85. DOI: 10.1016/j.memsci.2016.01.044

[40] Hou DY, Fan H, Jiang QL, Wang J, Zhang XH. Preparation and characterization of PVDF flat-sheet membranes for direct contact membrane distillation. Separation and Purification Technology. 2014;**135**:211-222. DOI: 10.1016/j.seppur.2014.08.023

[41] Hou DY, Dai GH, Fan H, Wang JN, Zhao CW, Huang HJ. Effects of calcium carbonate nano-particles on the

properties of PVDF/nonwoven fabric flat-sheet composite membranes for direct contact membrane distillation. Desalination. 2014;**347**:25-33. DOI: 10.1016/j.desal.2014.05.028

[42] Marino T, Russo F, Criscuoli A, Figoli A. TamiSolve (R) NxG as novel solvent for polymeric membrane preparation. Journal of Membrane Science. 2017;**542**:418-429. DOI: 10.1016/j.memsci.2017.08.038

[43] Tang YD, Li N, Liu AJ, Ding SK, Yi CH, Liu H. Effect of spinning conditions on the structure and performance of hydrophobic PVDF hollow fiber membranes for membrane distillation. Desalination. 2012;**287**: 326-339. DOI: 10.1016/j.desal. 2011.11.045

[44] Hou DY, Wang J, Sun XC, Ji ZG, Luan ZK. Preparation and properties of PVDF composite hollow fiber membranes for desalination through direct contact membrane distillation. Journal of Membrane Science. 2012;**405**:185-200. DOI: 10.1016/j. memsci.2012.03.008

[45] Hou DY, Wang J, Qu D, Luan ZK, Ren XJ. Fabrication and characterization of hydrophobic PVDF hollow fiber membranes for desalination through direct contact membrane distillation. Separation and Purification Technology. 2009;**69**(1):78-86. DOI: 10.1016/j. seppur.2009.06.026

[46] Hou DY, Wang J, Qu D, Luan ZK, Zhao CW, Ren XJ. Preparation of hydrophobic PVDF hollow fiber membranes for desalination through membrane distillation. Water Science and Technology. 2009;**59**(6):1219-1226. DOI: 10.2166/wst.2009.080

[47] Zhang H, Lu XL, Liu ZY, Ma Z, Wu S, Li ZD, et al. The unidirectional regulatory role of coagulation bath temperature on cross-section radius of the PVDF hollow-fiber membrane. Journal of Membrane Science. 2018;**550**:9-17. DOI: 10.1016/j. memsci.2017.12.059

[48] Li Y, Jin CL, Peng YL, An QF, Chen ZP, Zhang JC, et al. Fabrication of PVDF hollow fiber membranes via integrated phase separation for membrane distillation. Journal of the Taiwan Institute of Chemical Engineers. 2019;**95**:487-494. DOI: 10.1016/j. jtice.2018.08.036

[49] Garcia-Payo MC, Essalhi M, Khayet M. Effects of PVDF-HFP concentration on membrane distillation performance and structural morphology of hollow fiber membranes. Journal of Membrane Science. 2010;**347**(1-2): 209-219. DOI: 10.1016/j.memsci. 2009.10.026

[50] Garcia-Payo MC, Essalhi M, Khayet M. Preparation and characterization of PVDF-HFP copolymer hollow fiber membranes for membrane distillation. Desalination. 2009;**245**(1-3):469-473. DOI: 10.1016/j. desal.2009.02.010

[51] Khayet M, Cojocaru C, Essalhi M, Garcia-Payo MC, Arribas P, Garcia-Fernandez L. Hollow fiber spinning experimental design and analysis of defects for fabrication of optimized membranes for membrane distillation. Desalination. 2012;**287**:146-158. DOI: 10.1016/j.desal.2011.06.025

[52] Garcia-Payo MC, Essalhi M, Khayet M, Garcia-Fernandez L, Charfi K, Arafat H. Water desalination by membrane distillation using PVDF-HFP hollow fiber membranes. Membrane Water Treatment. 2010;**1**(3):215-230. DOI: 10.12989/ mwt.2010.1.3.215

[53] Alsalhy QF, Rashid KT, Ibrahim SS, Ghanim AH, Van der Bruggen B, Luis P, et al. Poly(vinylidene fluoride-co-hexafluoropropylene) (PVDF-co-HFP) hollow fiber

membranes prepared from PVDF-co-HFP/PEG-600Mw/DMAC solution for membrane distillation. Journal of Applied Polymer Science. 2013;**129**(6):3304-3313. DOI: 10.1002/app.39065

[54] Wang J, Zheng LB, Wu ZJ, Zhang Y, Zhang XH. Fabrication of hydrophobic flat sheet and hollow fiber membranes from PVDF and PVDF-CTFE for membrane distillation. Journal of Membrane Science. 2016;**497**:183-193. DOI: 10.1016/j.memsci.2015.09.024

[55] Meng SW, Mansouri J, Ye Y, Chen V. Effect of templating agents on the properties and membrane distillation performance of TiO_2-coated PVDF membranes. Journal of Membrane Science. 2014;**450**:48-59. DOI: 10.1016/j.memsci.2013.08.036

[56] Meng S, Ye Y, Mansouri J, Chen V. Fouling and crystallisation behaviour of superhydrophobic nano-composite PVDF membranes in direct contact membrane distillation. Journal of Membrane Science. 2014;**463**:102-112. DOI: 10.1016/j.memsci.2014.03.027

[57] Shaffer DL, Werber JR, Jaramillo H, Lin SH, Elimelech M. Forward osmosis: Where are we now? Desalination. 2015;**356**:271-284. DOI: 10.1016/j.desa12014.10.031

[58] Du H, Thompson A, Wang X. Osmotically Driven Membrane Processes: Approach, Development and Current Status. Rijeka: IntechOpen; 2018. DOI: 10.5772/intechopen.68607

[59] Camacho LM, Dumee L, Zhang JH, Li JD, Duke M, Gomez J, et al. Advances in membrane distillation for water desalination and purification applications. Water. 2013;**5**(1):94-196. DOI: 10.3390/w5010094

[60] Nave F, Kommalapati R, Thompson A. Introductory chapter: Osmotically driven membrane processes. In: Osmotically Driven Membrane Processes: Approach, Development and Current Status. Rijeka, Croatia: InTech; 2018. pp. 1-10. DOI: 10.5772/intechopen.72569

Chapter 5

Water Quality Ecological Risk Assessment with Sedimentological Approach

Limin Ma and Changxu Han

Abstract

The potential ecological risk index (ERI) is a useful diagnostic tool for water system assessment. It's based on sedimentology and combined with environmental chemistry and ecotoxicology. This chapter introduces the approach, including basic theory, calculation formula, evaluation criteria, and its parameters. Using a case study, the modification of the classification of the potential ecological risk is discussed. The water quality of the Liaohe River is assessed by the potential ecological risk index with the sedimentological approach. The sediments samples were collected from 19 sites and were analyzed for seven substances (Cd, As, Cu, Ni, Pb, Cr, and Zn) to assess the potential ecological risk. According to the results, Cd was found to be the main pollutant in the Liaohe River. The consequence of the monomial potential ecological risk factor E_r^i (mean) of each element is ranked as: Cd (93.39%) > As (3.13%) > Cu (1.26%) > Ni (0.97%) > Pb (0.70%) > Cr (0.34%) > Zn (0.22%). The ERI results (358.35) indicate the Liaohe River poses a very high potential ecological risk.

Keywords: water quality assessment, sedimentological method, Håkanson index, potential ecological risk index, methodologies

1. Introduction

The water and sediments are the main storage medium for pollutants in lake environments. The sediments adsorb various kinds of pollutants which could accumulate in sediments for a long time. When external conditions change, pollutants adsorbed in sediments may be released back into the water and taken up by organisms. Eventually, these pollutants may affect human health through the food chain. Therefore, how to assess the risk of the water system with contaminated sediments has become an important issue. If ecological risk assessment can be used as a diagnostic tool to evaluate the potential risks accurately, it is of great significance to pollution control [1, 2].

Until now, various approaches, which are based on the different perspectives of the chemical, biological and toxicological indices, have been proposed to assess the water quality ecological risk of the environment. For example, the enriched factor (EF) can evaluate the accumulation of elements in the sediment. It is calculated by comparing the concentration of the sample with the background value [3]. The geo-accumulation index (I_{geo}) assesses the risk by comparing the total concentration, the background value, and the background matrix correction factor of lithogenic effects is considered in it [4]. The pollution load index (PLI) is defined as the nth root of the product of the ratios between the concentration of each metal to the background values [5]. The sediment quality guidelines (SQGs) include threshold effect concentrations (TECs) and probable effect concentrations (PECs). Bioavailability is taken into account in this approach [6]. It is not adequate to assess the ecological risk by using only concentrations without factors of toxicity. The potential ecological risk index (ERI) posed by Swedish geochemist Lars Håkanson (The National Swedish Environment Protection Board, Water Quality Laboratory Uppsala) is based on the "abundance principle", "sink-effect", and "sensitivity factor" [7]. As a diagnostic tool for pollution control, the potential ecological risk index has been widely used since its development in the 1980s [8–10].

This chapter describes an approach to assess water quality risks using its basic theory, calculation formula, evaluation criteria, and parameters calculation. This approach combines environmental chemistry with ecotoxicology in order to assess the potential risks accurately. The approach integrates the concentration of substances with ecological effects, environmental effects, and toxicity. Furthermore, the model is used to explain in detail a water quality case study of the Liaohe River, China [11].

2. The potential ecological risk index

2.1 Theoretical hypothesis

Considering the different aspects that could affect ecological risk, Håkanson [7] made four hypotheses about the potential ecological risk index (ERI) value when he proposed the approach. They are:

1. The concentration requirement. The ERI value should increase as the pollutant contamination increases.

2. The number requirement. The ERI value should increase as the number of pollutant species increase.

3. The toxic factor requirement. Various substances have different toxicological effects. ERI value should differentiate between mildly, moderately and very toxic substances.

4. The sensitivity requirement. Various lakes and water systems do not have the same sensitivity to toxic substances.

2.2 Equations

Based on the above hypothesis, the potential ecological risk index is calculated by the following equations:

$$C_f^i = \frac{C_{0-1}^i}{C_n^i} \tag{1}$$

$$C_d = \sum_{i=1}^{n} C_f^i = \sum_{i=1}^{n} \frac{C_{0-1}^i}{C_n^i} \tag{2}$$

where C_f^i is the contamination factor of the substance i, C_{0-1}^i is the measured value of the substance i, C_n^i is the preindustrial reference value of the substance i, and C_d is the degree of contamination.

$$E_r^i = T_r^i \cdot C_f^i \tag{3}$$

$$ERI = \sum_{i=1}^{n} E_r^i = \sum_{i=1}^{n} T_r^i \cdot C_f^i \tag{4}$$

where E_r^i is the potential ecological risk factor for the given substance i, T_r^i is the "toxic-response" factor for the given substance i, and ERI is the potential ecological risk index for the basin/lake.

2.3 The parameters

2.3.1 The contamination factor C_f^i

To get the value of the contamination factor (C_f^i), more information needs to be known about the measured value of substance i (C_{0-1}^i) and the preindustrial reference value of substance i (C_n^i). In order to reflect the risk of the lake accurately, Håkanson proposed that "undisturbed" samples should be collected from accumulation areas in the lake targeting the 0–1 cm layer. Håkanson provides two methods to determine the accumulation areas for a given lake. The first method, the ETA-diagram (**Figure 1**), uses only the water depth and the effective fetch. The second method uses the water content of sediments (W_{0-1}). In this second method, researchers have to collect and analyze sediments to determine the bottom dynamic condition. The method requires 5 g wet sediment dried for 6 h at 105°C, then expressed as the water content as wet sediment. Accordingly, if the W_{0-1}>75%, it may mean the sediments are from an accumulation area.

In addition, Håkanson gives the types of contaminants that could be included in this contamination factor index. These contaminants include PCB, Hg, Cd, As, Cu, Pb, Cr, and Zn. Of course, it is possible to study other pollutants

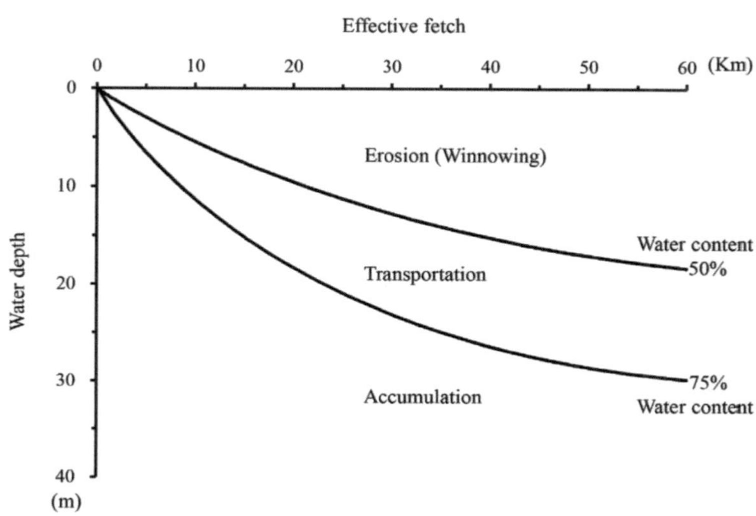

Figure 1.
The ETA-diagram [12].

(e.g., Ni, V, Mo, Co). Fe, Mn, and P are unsuitable as sediment parameters in this approach because their concentration is often influenced by physical or chemical processes in the sediments.

According to the contamination factor (C_f^i), single elements, C_f^i are classified as follows:

$C_f^i < 1$, low contamination factor;

$1 \leq C_f^i < 3$, moderate contamination factor;

$3 \leq C_f^i < 6$, considerable contamination factor;

$C_f^i \geq 6$, very high contamination factor.

For the preindustrial reference condition (C_n^i), Håkanson chose preindustrial background reference values as PCB = 0.01, Hg = 0.25, Cd = 1.0, As = 15, Cu = 50, Pb = 70, Cr = 90, and Zn = 175 (ppm). Different researchers [13–15] have selected other reference values for C_n^i, for example, the national standards and the background reference value.

2.3.2 The degree of contamination C_d

The degree of contamination value (C_d) is the sum of all C_f^i, which accounts for the total of the sediment pollution. C_f^i are classified as follows:

$C_d < 8$, low degree of contamination;
$8 \leq C_d < 16$, moderate degree of contamination;
$16 \leq C_d < 32$, considerable degree of contamination;
$C_d \geq 32$, very high degree of contamination.

The thresholds are determined by the number of substances. Eight substances were analyzed in Håkanson's research; therefore, the threshold is 8 for the low degree of contamination. C_d classification thresholds should be modified for different assessments. For example, if there are five substances analyzed in an assessment, then the threshold for the low degree of contamination should be 5.

2.3.3 The toxic factor St^i

In this risk index approach, the toxic factor (St^i) primarily provides two important pieces of information—the threat to man and the threat to the aquatic ecological system. Håkanson calculated the "toxic-response" factor based on "abundance principle" and "sink-effect". The potential biotoxicity of a metal element is inversely proportional to its abundance.

To evaluate the "abundance principle", the following methodology has been used:

1. The basic data for the evaluation is given in **Table 1**. It illustrates the abundance of various elements in igneous rocks, soils, fresh water, land plants, and land animals.

2. Relative abundance of elements in different media are shown in **Table 2**. The value of 1.0 is given to the element with the highest mean concentration in each media. For example, Zn has the highest value in land animals, so Zn should be given the value of 1.0.

Element	Igneous rocks	Soils	Freshwater	Land plants	Land animals
As	1.8	6.0	0.0004	0.2	≤0.2
Cd	0.2	0.06	<0.08	0.6	≤0.5
Cr	100	100	0.00018	0.23	0.075
Cu	55	20	0.01	14	2.4
Hg	0.08	0.03–0.8	0.00008	0.015	0.046
Pb	12.5	10	0.005	2.7	2.0
Zn	70	50	0.01	100	160

Table 1.
The abundance of various elements in different media ($\times 10^{-6}$) [16].

Order	Igneous rocks	Soils	Fresh water	Land plants	Land animals	$\sum_1^5$	$\sum_1^4$	Abundance number
1	1.0-Cr	1.0-Cr	1.0-Zn	1.0-Zn	1.0-Zn			
2	1.4-Zn	2.0-Zn	1.0-Cu	7.1-Cu	67-Cu			
3	1.8-Cu	5.0-Cu	2.0-Pb	37-Pb	80-Pb			
4	8.0-Pb	10-Pb	25-As	167-Cd	320-Cd			
5	56-As	17-As	31-Cd	435-Cr	800-As			
6	500-Cd	240-Hg	56-Cr	500-As	2130-Cr			
7	1250-Hg	1670-Cd	125-Hg	6670-Hg	3480-Hg			
Cr	1.0	1.0	56	435	2130*	2623	493.0	110.0
Zn	1.4	2.0*	1.0	1.0	1.0	6.4	4.4	1.0
Cu	1.8	5.0	1.0	7.1	67*	81.9	14.9	3.4
Pb	8.0	10	2.0	37	80*	137	57.0	13.0
As	56	17	25	500	800*	1398	598	140.0
Cd	500	1670*	31	167	320	2688	1018	230.0
Hg	1250	240	125	6670*	3480	11,765	5095	1160.0

**To avoid the inappropriate weight to the sum, the largest value for each element should be omitted.*

Table 2.
Relative abundance of elements in different media [17].

3. The "relative abundance" in each media is calculated by comparing the highest mean concentration with others in each media. For example, the value of Zn is 80 times higher than that of Pb in land animals, so Pb should be given 80. The results of relative abundance are given in **Table 2**.

4. The "abundance numbers" are determined by the sum of the five relative abundance numbers for each element. It is shown in the $\sum_1^5$ column. To balance the effect of extreme "abundance numbers" and to avoid the inappropriate weight to the "abundance numbers", the largest value marked "*" for each element should be omitted. The results of every element are given in the column marked $\sum_1^4$. In the end, the "abundance numbers" are obtained by division by the value of 4.4 (the value of Zn). For example, the "abundance

numbers" of Cr is obtained by dividing 493.0 (the sum of 1.0, 1.0, 56, and 435 in the line of Cr) by 4.4. The results of the "abundance numbers" are following: Zn < Cu < Pb < Cr < As < Cd < Hg.

5. The "corrected abundance numbers" are closely related to the toxicity coefficient, but it cannot represent "toxic-response" factor directly. Håkanson modified the "abundance numbers" by multiplying it by the "sink-factors", where the sink factor is determined as:

$$\text{Sink factor} = \frac{\text{Natural background concentration in fresh water}}{\text{Preindustrial reference value for lake sediments}}$$

Table 3 lists the data of natural background values for freshwater and preindustrial reference values. This results in the following "corrected abundance numbers": Zn = 57, Cr = 220, Cu = 680, Pb = 920, As = 3780, Cd = 46,000 and Hg = 371,200.

6. In order to match the dimensions of the contamination factors, first, divide all "corrected abundance numbers" by 57 (the value of Zn), then to take the square root of these figures, and then round off the values. This gives the following results: Zn = 1, Cr = 2, Cu = 5, P b = 5, As = 10, Cd = 30, and Hg = 80. The result of Hg is too high compared to Cd, therefore the toxic factor of Hg was determined as 40 by Håkanson. In addition, Håkanson hypothesized that the sedimentological toxic factor for PCB should be the same magnitude as that of Hg. Therefore, the St^i value for PCB was given 40. This gives the following St^i: Zn = 1, Cr = 2, Cu = 5, Pb = 5, As = 10, Cd = 30, Hg = 40, and PCB = 40.

2.3.4 The "toxic-response" factor T_r^i

It is well known that the sensitivity of organisms to the toxic substances is related to the biological characteristics of the aquatic systems [18]. This section describes sensitivity to toxic substances and how it varies from lake to lake. Håkanson uses the bioproduction index (BPI) value to represent the sensitivity. The BPI value is calculated by measuring the ignition loss (the IG value) and the nitrogen content (the N value) of sediment samples. The BPI value is defined as the nitrogen content on the regression line for IG = 10%. The nitrogen content is

Element	Background concentration in fresh water	Preindustrial reference value for lake sediments	Sink factor (10^{-3})	Abundance number	Corrected abundance numbers
Cr	0.2	90	2	110.0	220
Zn	10	175	57	1.0	57
Cu	10	50	200	3.4	680
Pb	5	70	71	13.0	920
As	0.4	15	27	140.0	3780
Cd	0.2	1	200	230.0	46,000
Hg	0.08	0.25	320	1160.0	371,200

Table 3.
Sink factors of elements [16].

determined using the standard Kjeldahl method [19]. The IG value is the ignition loss of dried sediment samples (550°C for 1 h). The N value and IG value are given in mg/g and % ds (ds = dry substance), respectively. After Håkanson's analysis, the relationships between the BPI value and St^i are the following (**Table 4**).

2.3.5 The monomial potential ecological risk factor E_r^i

The monomial potential ecological risk factor (E_r^i) is used to express the potential ecological risk for a substance. E_r^i values are classified as follows:

$E_r^i < 40$, low potential ecological risk;
$40 \le E_r^i < 80$, moderate potential ecological risk;
$80 \le E_r^i < 160$, considerable potential ecological risk;
$160 \le E_r^i < 320$, high potential ecological risk;
$E_r^i \ge 320$, very high ecological risk.

It is should be noted that the thresholds of low potential ecological risk are determined by the largest T_r^i value of substances. This means that even though there is no contamination ($C_f^i = 1$), the E_r^i can reach a value of 40 [20].

2.3.6 The comprehensive potential ecological index ERI

The comprehensive potential ecological risk index (ERI) is the sum of all E_r^i values which is used to express the potential ecological risk for a given aquatic system. ERI values are classified as follows:

ERI < 150, low potential ecological risk for the water system.
$150 \le$ ERI < 300, moderate potential ecological risk for the water system.
$300 \le$ ERI < 600, considerable potential ecological risk for the water system.
ERI ≥ 600, very high ecological risk for the water system.

The thresholds of C_d and E_r^i values are determined by the number and type of contaminants. The thresholds of ERI value are determined similarly. ERI values are determined by the sum of all the T_r^i values of every substance in an assessment. It could consider that there is a reference lake in which each substance's C_f^i

Substance	St^i value	T_r^i value
PCB	40	$40 \cdot BPI/5$
Hg	40	$40 \cdot 5/BPI$
Cd	30	$30 \cdot \sqrt{5/\sqrt{BPI}}$
As	10	10
Cu	5	$5 \cdot \sqrt{5/\sqrt{BPI}}$
Pb	5	$5 \cdot \sqrt{5/\sqrt{BPI}}$
Cr	2	$2 \cdot \sqrt{5/\sqrt{BPI}}$
Zn	1	$1 \cdot \sqrt{5/\sqrt{BPI}}$

Table 4.
The St^i and T_r^i of elements [7].

value = 1.0, BPI value = 5.0. This means that there is no contamination in the reference lake. The data from one's samples would be compared with the reference lake. The ERI classification thresholds are modified for different assessments. For example, if there were eight substances analyzed in Håkanson's research and the sum of all the T_r^i values is 155, the thresholds of the first level could be 150. Moreover, Håkanson ignores the influence of BPI value on the T_r^i value because of the C_f^i value is 1.0. Therefore, he regards the sum of St^i value as the threshold.

3. Case application

This section illustrates the potential ecological risk index by using a case study. The data for the ERI values is taken from [11]. The main steps for creating a potential ecological risk index to assess the Liaohe River system are:

1. Determine the substances of interest (As, Cd, Cr, Cu, Ni, Pb, and Zn) in the study area (the Liaohe River);

2. Determine the accumulation areas for the river and collect the samples from the 0–1 cm layer in the sediments;

3. Calculate or look up the St^i value;

4. Measure the IG value and N value to calculate the BPI and T_r^i value; and,

5. Calculate the potential ecological risk to assess the water quality.

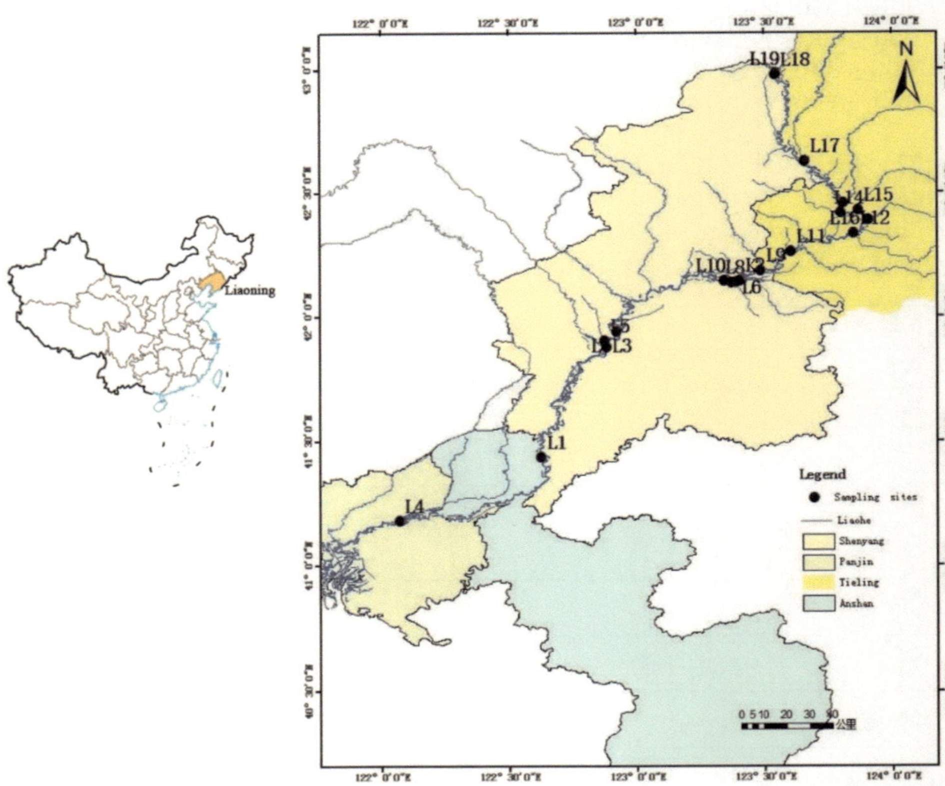

Figure 2.
The location of sampling sites along the Liaohe River protected area [11].

3.1 Description of the study area

The Liaohe River is located in the south of northeast China (**Figure 2**). It is one of the seven major rivers in China. As an important aquatic ecosystem, it plays an important role in the local economic and social development. Because of anthropogenic activities, the pollution of the Liaohe River is becoming a more serious problem. The Liaohe River has become one of the most polluted rivers in China. Therefore, it is significant to assess the quality of the Liaohe River [21].

3.2 Data collection and processing

Nineteen superficial sediment samples were collected along the Liaohe River protected area. At each site, three surface sediments were collected and placed into polyethylene bags and sealed. An Inductively Coupled Plasma Optical Emission Spectrometer (ICP-OES) was applied for the determination of heavy metals (As, Cd, Cr, Cu, Ni, Pb, and Zn). The details are found in [11].

				C_f^i				C_d
	Cd	**As**	**Cu**	**Pb**	**Ni**	**Cr**	**Zn**	
L1	3.70	0.70	0.40	0.38	0.48	0.67	0.42	6.75
L2	4.94	1.28	0.57	0.43	0.62	0.53	0.49	8.86
L3	19.75	0.82	0.78	0.47	0.56	0.44	1.15	23.97
L4	20.06	0.44	0.57	0.37	0.43	0.37	1.07	23.31
L5	20.99	0.57	0.87	0.47	0.66	0.32	1.22	25.09
L6	19.44	0.81	1.08	0.57	0.79	0.62	1.47	24.78
L7	18.21	0.39	0.66	0.34	0.36	0.36	0.81	21.13
L8	5.87	1.33	0.63	0.44	0.64	0.84	1.02	10.77
L9	5.56	1.38	2.08	0.59	1.38	0.89	1.00	12.86
L10	6.79	1.64	1.49	0.57	1.31	0.99	0.84	13.63
L11	5.56	1.31	1.01	0.51	0.82	0.83	0.79	10.83
L12	5.25	1.27	0.69	0.46	0.71	0.69	0.67	9.72
L13	6.48	1.07	1.17	0.50	0.90	0.84	0.71	11.67
L14	4.35	0.91	0.80	0.38	0.59	0.49	0.42	7.94
L15	9.91	1.16	0.60	0.54	0.47	0.47	0.53	13.67
L16	5.83	1.11	0.57	0.47	0.41	0.46	0.50	9.34
L17	13.43	1.59	1.45	0.70	0.76	0.63	0.57	19.13
L18	25.00	2.00	0.57	0.72	0.47	0.49	0.58	29.83
L19	10.83	1.57	1.14	0.60	0.79	0.63	0.76	16.32
Min	3.70	0.39	0.40	0.34	0.36	0.32	0.42	6.75
Max	25.00	2.00	2.08	0.72	1.38	0.99	1.47	29.83
Mean	11.16	1.12	0.90	0.50	0.69	0.61	0.79	15.77
Reference lake ("unpolluted")	1.00	1.00	1.00	1.00	1.00	1.00	1.00	7.00

Table 5.
Contamination factors (C_f^i) of different elements detected in sediments.

3.3 Methods

The potential ecological risk index is used to assess the ecological risk of the Liaohe River. The computational formula was shown as Eqs. (1)–(4). The T_r^i for Cd, As, Cu, Pb, Ni, Cr, and Zn are 30, 10, 5, 5, 5, 2, and 1, respectively [7, 22].

3.4 Results

3.4.1 The degree of contamination C_d

Table 5 shows the contamination factor C_f^i of the substances in the sediments from the Liaohe River. In Håkanson's research, seven metals (Hg, Cd, As, Cu, Pb, Cr, and Zn) and one organic pollutant (PCBs) were considered. However, in this study, there are only seven metals considered. Therefore, the C_d classification thresholds are modified. According to Håkanson's approach, the threshold for the "low degree of contamination" is 7, corresponding to the number of substances (7). The classification of C_f^i and C_d are classified in **Table 6**.

Table 5 shows that the C_f^i values of sampling sites range from 0.32 to 25.00. The average C_f^i value of each element and the percentage of that in C_d are in the following

Threshold	Modified threshold	Degree of risk
$C_f^i < 1$	/	Low
$1 \leq C_f^i < 3$	/	Moderate
$3 \leq C_f^i < 6$	/	Considerable
$C_f^i \geq 6$	/	Very high
$C_d < 8$	$C_d < 7$	Low
$8 \leq C_d < 16$	$7 \leq C_d < 14$	Moderate
$16 \leq C_d < 32$	$14 \leq C_d < 28$	Considerable
$C_d \geq 32$	$C_d \geq 28$	Very high
$E_r^i < 40$	$E_r^i < 30$	Low
$40 \leq E_r^i < 80$	$30 \leq E_r^i < 60$	Moderate
$80 \leq E_r^i < 160$	$60 \leq E_r^i < 120$	Considerable
$160 \leq E_r^i < 320$	$120 \leq E_r^i < 240$	High
$E_r^i \geq 320$	$E_r^i \geq 240$	Very high
ERI < 150	ERI < 60	Low
150 ≤ ERI < 300	60 ≤ ERI < 120	Moderate
300 ≤ ERI < 600	120 ≤ ERI < 240	Considerable
ERI ≥ 600	ERI ≥ 240	Very high

Table 6.
Classification of the potential ecological risk.

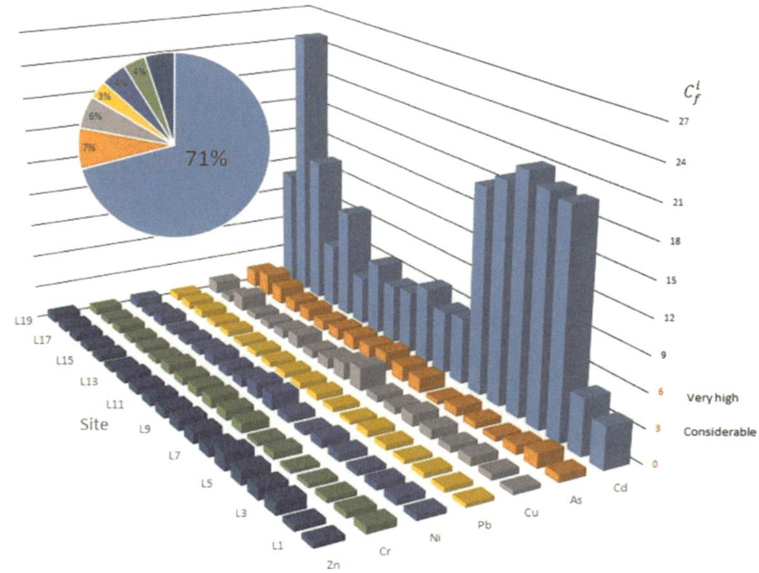

Figure 3.
Contamination factors (C_f^i) of different elements detected in sediments.

	Elements (St^i value)							$ERI = \sum_{i=1}^{7} E_r^i$
	Cd	As	Cu	Pb	Ni	Cr	Zn	
	30	10	5	5	5	2	1	
L1	111.11	6.97	1.99	1.90	2.41	1.34	0.42	126.14
L2	148.15	12.84	2.84	2.15	3.12	1.05	0.49	170.64
L3	592.59	8.22	3.92	2.34	2.81	0.87	1.15	611.90
L4	601.85	4.39	2.85	1.87	2.17	0.73	1.07	614.93
L5	629.63	5.68	4.33	2.36	3.29	0.63	1.22	647.14
L6	583.33	8.07	5.38	2.86	3.96	1.24	1.47	606.31
L7	546.30	3.90	3.31	1.71	1.80	0.71	0.81	558.54
L8	175.95	13.33	3.15	2.20	3.21	1.67	1.02	200.53
L9	166.67	13.79	10.38	2.95	6.89	1.77	1.00	203.45
L10	203.70	16.36	7.44	2.87	6.56	1.98	0.84	239.75
L11	166.67	13.14	5.03	2.57	4.11	1.65	0.79	193.96
L12	157.41	12.69	3.43	2.28	3.53	1.38	0.67	181.39
L13	194.44	10.68	5.86	2.50	4.52	1.67	0.71	220.38
L14	130.56	9.13	3.99	1.90	2.93	0.98	0.42	149.91
L15	297.22	11.56	2.98	2.70	2.36	0.94	0.53	318.29
L16	175.00	11.06	2.85	2.33	2.04	0.92	0.50	194.70
L17	402.78	15.91	7.26	3.50	3.79	1.26	0.57	435.07
L18	750.00	20.03	2.85	3.60	2.34	0.98	0.58	780.38
L19	325.00	15.65	5.68	3.01	3.96	1.26	0.76	355.32
Min	111.11	3.90	1.99	1.71	1.80	0.63	0.42	126.14

	Elements (St^i value)							
	Cd	As	Cu	Pb	Ni	Cr	Zn	$ERI = \sum_{i=1}^{7} E_r^i$
	30	10	5	5	5	2	1	
Max	750.00	20.03	10.38	3.60	6.89	1.98	1.47	780.38
Mean	334.65	11.23	4.50	2.51	3.46	1.21	0.79	358.35
Reference lake ("unpolluted")	30	10	5	5	5	2	1	58

Table 7.
The potential ecological risk factor (E_r^i) of different elements detected in sediments [11].

order: Cd (70.74%) > As (7.12%) > Cu (5.71%) > Zn (5.01%) > Ni (4.39%) > Cr (3.84%) > Pb (3.18%). Every C_f^i value of Pb and Cr is less than 1.0. For the average C_f^i value, Cd and As have a very high and moderate contamination factor, respectively. Whereas, Cu, Zn, Ni, Cr, and Pb have low contamination factors.

The resulting C_d values of each sample site ranged from 6.75 to 29.83. According to the category of C_d (**Table 6**), only sample L1 has the low degree of contamination. Ten sampling sites are classified as moderate and 7 sampling sites as having high contamination factors, sample L19 is classified into very high contamination factor. **Figure 3** clearly shows that Cd has the highest contamination factor. That means the Liaohe River is dominated by the pollution of one element—Cadium.

3.4.2 The potential ecological risk E_r^i and ERI

If the classification thresholds of C_d are modified, the E_r^i and ERI should also be modified. The first level of E_r^i is fixed by the T_r^i value of the most toxic element. This means that the results of the given water body are compared with a reference lake which has no contamination (C_f^i = 1). Similarly, the first level of ERI is fixed by the sum of T_r^i value of all the elements.

In the Liaohe River case study, the most toxic element is Cd and the T_r^i of Cd is 30. Therefore, the classification threshold of E_r^i is 30. The sum of T_r^i of all elements

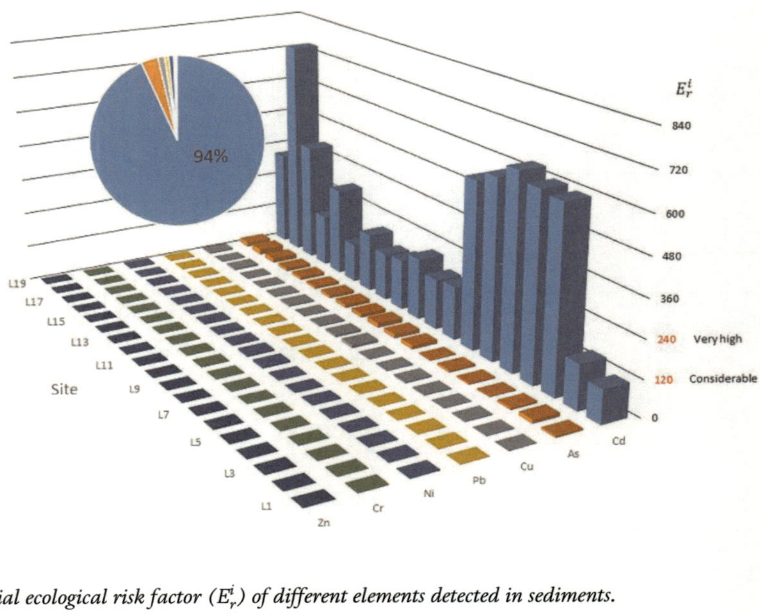

Figure 4.
The potential ecological risk factor (E_r^i) of different elements detected in sediments.

is 58, so the classification threshold of ERI could be 60. The classification of E_r^i and ERI are classified in **Table 6**.

Table 7 illustrates the potential ecological risks of the heavy metals in the sediments from the Liaohe River. The E_r^i values of sampling sites range from 0.42 to 750.00. The consequence of E_r^i (mean) of the 7 heavy metals are ranked as: Cd (93.39%) > As (3.13%) > Cu (1.26%) > Ni (0.97%) > Pb (0.70%) > Cr (0.34%) > Zn (0.22%). The E_r^i value of As, Cu, Pb, Ni, Cr, and Zn are all below 30. According to the category of E_r^i (**Table 6**), these six heavy metals have a low potential ecological risk. Cd at L1 posed a considerable potential ecological risk (111.11), while at other sampling sites, it shows high or very high potential ecological risk. The very highest E_r^i value is observed for Cd (750.00) at L18, indicates extremely severe pollution. The ERI values for the sampling sites range from 126.14 to 780.38. According to the listing of the ERI values (**Table 6**), the lowest ERI value for site L1 is over 120; therefore, all the sampling sites all have the considerable or very high potential ecological risk. The mean value of ERI (358.35) for the sediments in the Liaohe River indicates very high potential ecological risk (**Figure 4**).

4. Discussion

The Liaohe River is used as a case study to illustrate this approach. The investigation of seven heavy metals (Cd, As, Cu, Ni, Pb, Cr, and Zn) in the sediments suggest that the Liaohe River is dominated by the pollution of Cd which contributes around 94% potential ecological risk. The E_r^i means of the remaining sites are ranked as: Cd (93.39%) > As (3.13%) > Cu (1.26%) > Ni (0.97%) > Pb (0.70%) > Cr (0.34%) > Zn (0.22%). All elements except cadmium have low potential ecological risk. According to the ERI results, due to the serious pollution of cadmium, all the sampling sites have the considerable or very high potential ecological risk. Thus, it is important to control the pollution of cadmium. This study assesses the risk of Liaohe River by the modified risk classification criterion. Therefore, the results are different from [11], the risks assessed by this study are more serious. It is worth discussing how to use the risk classification criterion. This study suggests using the modified risk classification criteria.

Because of the "toxic-response" factor, compared with other approaches, the potential ecological risk index can distinguish the differences among substances and aquatic systems. Therefore, this approach has outstanding advantages to assess the risk of water system as a widely used approach which can provide a better overall ecological risk to the aquatic system. However, two main problems are neglected in the application of this method. (1) T_r^i is replaced by St^i. More attention should be given to the BPI value. Different aquatic systems have different sensitivities to toxic substances. According to Eq. (3) and **Table 4**, the effect of BPI value on the results depends on the degree of contamination of the aquatic system. If the pollution of the study aquatic system is serious, the BPI value will have large effect on the index calculation. Ecological risks can be evaluated more accurately by measuring the BPI value of the study aquatic system. (2) According to Håkanson's research [7, 23], the classification thresholds should be modified for different assessments. In this chapter, a reasonable suggestion for modification is suggested as well as applied. For C_d, the threshold for the "low risk" is modified by the number of substances. For E_r^i, the threshold for the "low risk" is modified by the T_r^i value of the most toxic element. For ERI, the threshold for the "low risk" is modified by the sum of T_r^i of all elements. There are still other problems deserve researchers concerns in the application of this approach, for example, the determination of accumulation areas in the aquatic system and calculation of St^i value. This study provides detail information for the potential ecological risk index and discusses several problems of the approach. And it is helpful for researchers to assess the ecological risk of aquatic system by this approach.

Acknowledgements

This work was supported by the Key Program of China (2018YFC1803103) and National Natural Science Foundation of China (No. 21377098). We also thank the anonymous reviewers and the editors for their comments which improved the manuscript.

Author details

Limin Ma[1,2,3*] and Changxu Han[1]

1 State Key Laboratory of Pollution Control and Resources Reuse, Tongji University, Shanghai, China

2 Key Laboratory of Yangzi River Water Environment, Ministry of Education, Tongji University, Shanghai, China

3 Shanghai Institute of Pollution Control and Ecological Security, Shanghai, People's Republic of China

*Address all correspondence to: lmma@tongji.edu.cn

References

[1] Li F, Huang J, Zeng G, et al. Spatial risk assessment and sources identification of heavy metals in surface sediments from the Dongting Lake, Middle China. Journal of Geochemical Exploration. 2013;**132**:75-83

[2] Wu Y, Xu Y, Zhang J, et al. Evaluation of ecological risk and primary empirical research on heavy metals in polluted soil over Xiaoqinling gold mining region, Shaanxi, China. Transactions of the Nonferrous Metals Society of China. 2010;**20**:688-694

[3] Taylor SR. Abundance of chemical elements in the continental crust: A new table. Geochimica et Cosmochimica Acta. 1964;**28**:1273-1285

[4] Müller G. Index of geo-accumulation in sediments of the Rhine River. Geochemical Journal. 1969;**2**:108-118

[5] Tomlinson DL, Wilson JG, Harris CR, et al. Problems in the assessment of heavy-metal levels in estuaries and the formation of a pollution index. Helgoländer Wissenschaftliche Meeresuntersuchungen. 1980;**33**: 566-575

[6] MacDonald DD, Ingersoll CG, Berger TA. Development and evaluation of consensus-based sediment quality guidelines for freshwater ecosystems. Archives of Environmental Contamination and Toxicology. 2000; **39**:20-31

[7] Håkanson L. An ecological risk index for aquatic pollution control. A sedimentological approach. Water Research. 1980;**14**:975-1001

[8] Sun C, Zhang Z, Cao H, et al. Concentrations, speciation, and ecological risk of heavy metals in the sediment of the Songhua River in an urban area with petrochemical industries. Chemosphere. 2019;**219**: 538-545

[9] Zhang M, He P, Qiao G, et al. Heavy metal contamination assessment of surface sediments of the Subei Shoal, China: Spatial distribution, source apportionment and ecological risk. Chemosphere. 2019;**223**:211-222

[10] Zhang Z, Lu Y, Li H, et al. Assessment of heavy metal contamination, distribution and source identification in the sediments from the Zijiang River, China. Science of the Total Environment. 2018;**645**:235-243

[11] Ke X, Gui S, Huang H, et al. Ecological risk assessment and source identification for heavy metals in surface sediment from the Liaohe River protected area, China. Chemosphere. 2017;**175**:473-481

[12] Håkanson L. The influence of wind, fetch, and water depth on the distribution of sediments in Lake Vänern, Sweden. Canadian Journal of Earth Sciences. 1977;**14**:397-412

[13] Liu R, Bao K, Yao S, et al. Ecological risk assessment and distribution of potentially harmful trace elements in lake sediments of Songnen Plain, NE China. Ecotoxicology and Environmental Safety. 2018;**163**:117-124

[14] Ram SS, Aich A, Sengupta P, et al. Assessment of trace metal contamination of wetland sediments from eastern and western coastal region of India dominated with mangrove forest. Chemosphere. 2018;**211**: 1113-1122

[15] Tran TAM, Leermakers M, Hoang TL, et al. Metals and arsenic in sediment and fish from Cau Hai lagoon in Vietnam: Ecological and human health risks. Chemosphere. 2018;**210**: 175-182

[16] Bowen HJM. Trace Elements in Biochemistry. London/New York: Academic Press; 1966

[17] Forstner U, Müller G. Schwermetalle in Flüssen und Seen als Ausdruck der Umweltverschmutzung. Berlin Heidelberg: Springer; 1974

[18] Dauvalter VA, Kashulin NA. Assessment of the ecological state of the Arctic Freshwater system based on concentrations of heavy metals in the bottom sediments. Geochemistry International. 2018;**56**:842-856

[19] Kjeldahl J. Neue methode zur bestimmung des stickstoffs in organischen körpern. Fresenius' Zeitschrift für Analytische Chemie. 1883;**22**:366-382

[20] Carballeira JAFA. Evaluation of contamination, by different elements, in terrestrial mosses. Archives of Environmental Contamination and Toxicology. 2001;**40**:461-468

[21] Ke X, Gao L, Huang H, et al. Toxicity identification evaluation of sediments in Liaohe River. Marine Pollution Bulletin. 2015;**93**:259-265

[22] Xu Z, Ni S, Tuo X, et al. Calculation of heavy metals' toxicity coefficient in the evaluation of potential ecological risk index. Environmental Sciences. 2008;**2**:112-115

[23] Liu C, Wang Z, He G, et al. Evaluation on the potential ecological risk for the river mouths around Bohai Bay. Research of Environmental Sciences. 2002;**15**:33-37

Section 3

Water Quality Assessments

Chapter 6

Jewels across the Landscape: Monitoring and Assessing the Quality of Lakes and Reservoirs in the United States

David V. Peck, Steven G. Paulsen, Philip R. Kaufmann and Alan T. Herlihy

Abstract

An early naturalist described lakes as "jewels" across the landscape and indeed they were...at the end of the nineteenth century. As we settled the country and began to utilize the lake resource for our needs, things changed. Additionally, our needs for water brought about the construction of impoundments from ice ponds to small stock ponds up to mainstem impoundments along our major rivers. The lake resource in the United States now includes natural lakes in our northern tier of states, unique physiographic regions such as Florida and the Sand Hills of Nebraska, and the mountainous regions, and impoundments scattered across the entire landscape. In this chapter, we will describe efforts by an unique partnership between the individual states and tribal nations of the USA and the US Environmental Protection Agency to monitor and assess these systems. These efforts go beyond single water quality (chemistry) issues and include assessments targeting the goal of the Clean Water Act, namely, restoring, maintaining, and protecting the chemical, physical, and biological integrity of the nation's lakes and reservoirs.

Keywords: lakes, reservoirs, monitoring, assessment, National Lakes Assessment, United States, ecological indicators, survey design, National Aquatic Resource Assessments, water quality, trophic state, biological integrity, lakeshore habitat, Clean Water Act

1. Introduction

The United States' love affair with lakes dates back a long way. In 1896, MacGonigle [1] described lakes in central Florida this way: "Dotting the landscape, like jewels of crystal in a field of green, are numberless lakes, varying in size from a gem-like lakelet to the broad expanse of Okeechobee". Many states, in particular Vermont, New York, Maine, Michigan, Wisconsin, and Minnesota, have extensive histories and ties with their lakes. In this chapter, we discuss why lake monitoring is important, and what are the essential characteristics of the U.S. National Lakes Assessment (NLA) that allow us to rigorously characterize the status of this precious lake resource and track how the status is changing over time.

2. Background

The US Clean Water Act (CWA) of 1972 [2] expresses the national desire to restore and maintain the physical, chemical, and biological integrity of USA waters and requires that information on status and trends be reported every 2 years by the states. Different States vary greatly in their monitoring focus and approaches. It has long been recognized that these reports cannot be combined to create a coherent picture of the degree to which lakes in the USA meet the goals of the CWA [3–9]. Looking back at the history of national lake assessments, it is clear that our focus in assessing lake condition has shifted over time as each new threat to lake quality emerged. In the 1960s–1970s, our focus was the "cultural eutrophication" of lakes, that is, the nutrient enrichment of lakes through human activities, via point or nonpoint sources of organic and inorganic nutrients. This enrichment led to everything from "unsightly" algal growth to health problems associated with recreational contact. When extreme, these algal blooms eventually led to low dissolved oxygen levels as the algae died and decayed. The low dissolved oxygen ultimately led to die-off of sensitive fish communities in many lakes. These concerns about eutrophication led to the first ever national lake survey in the USA, the National Eutrophication Survey (NES) [10]. The survey focused on lakes near population centers that were likely subjected to point-source release of nutrients or oxygen demanding compounds. Over 800 lakes suspected of having problems were sampled during this survey using a targeted approach. Ultimately, these concerns led to the funding of the Clean Lakes Program, a Congressionally funded program managed by the fledgling Environmental Protection Agency to provide states and communities with funding to solve specific problems with individual lakes.

The concern about eutrophication and desire to engage the public through citizen monitoring continued into the 1990s. In 1994, the National Secchi Dip-In program was implemented. The Dip-In is a volunteer effort in which citizens from various localities send in their Secchi Depth readings (a measure of lake water clarity) for lakes of interest during a particular week during the summer. This event continues under the sponsorship of the North American Lake Management Society [11].

The 1980s saw increasing concerns about releases of nitrogen and sulfur compounds into the atmosphere and the deposition of these acidic compounds onto lakes and stream watersheds in poorly buffered landscapes. When inquiring into the extent of the problem at the time, William Ruckelshaus, the EPA Administrator at the time, was rumored to have said something along the lines of "What do you mean you don't know how many acid lakes there are?" A definitive answer to this question was not possible at that time for several reasons, including the uncertainty in extrapolating results from site-specific studies to regional or national populations of lakes [12]. These concerns, in Europe and North America, particularly in highly visible regions like the Adirondacks, eventually led to the implementation of the National Acidic Precipitation Assessment Program (NAPAP). Key projects within NAPAP were the National Surface Water Surveys (NSWS), probability-based surveys of lakes (and streams) that set out to document how many acidic lakes and streams there were in the U.S. and how these systems might be changing in response to the 1990 Clean Air Act Amendments [12–14].

Following the completion of the initial NAPAP-sponsored surveys, EPA began to ask whether there might be a better, more consistent approach to directly address the CWA objectives for assessing the condition of lakes and other important ecological resources rather than mounting new surveys for each new problem that arose. The Environmental Monitoring and Assessment Program (EMAP) was a research program designed to develop this approach [6, 15] with a focus on CWA objectives. These research efforts culminated, for lakes, in the implementation and completion of the EMAP Northeastern Lakes Regional Demonstration Project conducted from 1991 to 1995 in the New England states, New Jersey, and New York [16–18].

As the EMAP research efforts on lake, stream, river, wetland, and estuary monitoring demonstrated their potential effectiveness, the US Office of Management and Budget (OMB) directed the EPA Office of Water to partner with the individual states of the USA to implement the EMAP concepts on a national scale for all waterbody types under the National Aquatic Resources Surveys (NARS). The first National Lake Assessment (NLA) was conducted in 2007 with recurrent surveys in 2012 and 2017 and planned surveys for every 5 years following. The description below outlines the conceptual and practical basis for the lakes monitoring efforts are taking place as part of the NLA.

3. Conceptual approach

Three aspects of the NLA make up the overall conceptual approach – the selection of indicators, the approach (survey design) for selecting sites to sample and making inferences to all lakes, and the strategy (response design) for acquiring data at each site for all indicators [6, 19]. This conceptual approach ensures that the NLA will address the main goal of the CWA as well as address the five big questions most frequently asked by the public:

1. Is there a problem with the condition of lakes?

2. How big is the problem?

3. Is the problem widespread or localized in hot spots?

4. Is the problem getting better or worse?

5. What is causing the problem?

Past surveys of lakes have pursued individual stressors or anthropogenic problems and measured them, for example, the National Eutrophication Survey focused on nutrients, phosphorus in particular, and the National Surface Water Surveys (NSWS) under NAPAP focused on acidification. The NLA, under NARS, is intended to have a broader perspective by using a variety of indicators to examine the overall health of lakes and ranking the importance of individual anthropogenic stressors.

This perspective drove the NLA to focus on indicators related to the attribute of "biological integrity" referenced in the CWA [2] to describe "condition" of lakes. In addition, indicators of "physical integrity" and "chemical integrity" describe the relative importance of human-mediated disturbances impacting lake condition.

The survey design plays a critical role in the overall approach within the NARS and the NLA. Frequently, surveys are developed with little attention to the final statements that are intended to be made from the data. The National Eutrophication Survey, for example, was based on a targeted judgment sample of 817 lakes potentially influenced by nutrient inputs from domestic wastewater treatment plants. Without statistically representative site selection, the only conclusions that could be reliably made from the data were about those 817 specific lakes. The Great Secchi Dip-In acquires data from thousands of lakes each year. The results provide important information about those lakes being monitored, but because the lakes selected for sampling are chosen by those submitting the data, the results are not necessarily representative of the total lake population (e.g., see [20]). The lake surveys conducted as part of the National Surface Water Surveys (NSWS) used a statistical design restricted to acid-sensitive regions (rather than the whole country) that allowed inference to be made from the sampled lakes to the greater population of lakes they represented in those defined areas. Because the focus was on acidification and acid deposition, the selection of lakes was understandably limited to lakes in regions of the country that had poor buffering capacity in the soils. Therefore, these lakes were potentially sensitive to acidification from acids in atmospheric deposition. By contrast, the NLA is the first national survey that focuses on all waterbodies in the conterminous U.S. meeting the definition of a lake (both natural and man-made) and employs a survey design that ensures that inferences can be made to that full "target" population of lakes [21]. More details of the NLA survey design are provided in following sections of this chapter.

The final aspect of the conceptual approach for indicators or measurements is the "response design," that is, when the crews get to specific lakes, where and how do they collect samples or measurements for the various indicators? This will be described in more detail below.

3.1 Indicators

Indicators used in the NLA are selected to assess status related to trophic state, water quality, the condition of biological assemblages, physical habitat condition, and human use (**Table 1**). The set of selected indicators are intended to be most appropriate for the assessment of lake condition at regional and national scales. Indicators range from direct measurements of specific variables to more complex indices representing biological or physical habitat condition.

3.2 Survey design

The target population (i.e., the set of lakes about which inferences are to be made) for the NLA includes all natural lakes and ponds, reservoirs, and man-made ponds within the conterminous USA (i.e., the "lower" 48 states) that are greater than 1 hectare (ha) in surface area, are permanent waterbodies, have an estimated maximum depth greater than 1 m, and have more than 1000 m^2 of open water on the day of sampling. An early decision was made to sample lakes as a finite resource and provide estimates of "lake number" and "proportion of lake number" rather than as "lake area" (although areal estimates can also be made with the NLA data). The NLA design requires some level of stratification or unequal sampling probability to accommodate regional variation in the abundance of lakes, and the preponderance of small lakes [22, 23]. A simple random sample will be dominated by small lakes (less than 4 ha), and the bulk of lakes sampled will be in the Upper Midwest where lakes are most abundant. Because of the desire to make both national and regional estimates, care is taken to spread the sample across the conterminous USA and across the size range of lakes available. For regional coverage, variable selection probabilities are set to ensure the ability to describe conditions in all 10 EPA Regions [24], 9 aggregated NARS ecoregions (**Figure 1**) [25] and roughly 15 hydrologic basins. Variable selection probabilities are also set to ensure that the NLA samples are spread across the size range of lakes so that small lakes do not dominate the sample. Samples are currently allocated among 5 lake surface area categories: 1–4, 4–10, 10–20, 20–50 ha, and greater than 50 ha. Each site sampled receives a "weight" inversely proportional to its probability of inclusion in the sample. The weights are then used to make the inferences from

Indicator and rationale	Sample location
Zooplankton assemblage: important element of the food web; responds to stressors such as nutrient enrichment and acidification	Collected from the upper portion of the water column at the open-water site. Organisms were usually identified to genus and an multimetric index was developed based on life history characteristics and tolerance to environmental conditions
Trophic state (chlorophyll *a*): responsive to nutrient enrichment and can be associated with risk of harmful algal blooms	A trophic state index was calculated based on measured chlorophyll *a* concentration
Benthic macroinvertebrate assemblage: responsive to a variety of stressors and can integrate exposure to current and recent past levels	Kicknet samples collected from the lake bottom at 10 shoreline locations and combined into a single composite sample for each lake. Organisms were usually identified to genus and a multimetric index was developed based on life history characteristics and tolerance to environmental conditions
Total phosphorus: important nutrient affecting trophic state and algal community structure	Collected from a vertically integrated sample of the upper water column at the open-water site. Measured concentrations were compared to benchmarks
Total nitrogen: important nutrient affecting trophic state and algal community structure	Collected from a vertically integrated sample of the upper water column at the open-water site. Measured concentrations were compared to benchmarks
Dissolved oxygen: low levels can result from nutrient enrichment and lead to loss of biota	In situ measurements were collected from the entire water column at the open-water site. The mean value of measurements from the top 2 m of the profile was calculated and compared to benchmarks
Acidification (acid neutralizing capacity—ANC): indicates potential exposure to episodic or chronic acidification, which can affect structure and composition of algal, zooplankton, and fish assemblages	ANC (corrected for DOC) measured from a vertically integrated sample of the upper water column at the open-water site. Measured concentrations were compared to benchmarks
Lake habitat complexity: indicates effects of human activities on the complexity of cover features in the riparian, shoreline, and littoral zones. Supports diversity of biotic assemblages such as fish, benthic invertebrates, and birds	Observations were recorded from 10 shoreline locations around each lake. Observed indicator values were compared with lake-specific expected values based on natural controlling factors within each region. Condition determinations were based on magnitude of deviations from expected values
Shallow water habitat: indicates effects of human activities on or near lakeshores on the complexity of littoral cover features that support biota	Same as for lake habitat complexity
Lakeshore disturbance: indicates types and potential severity of human activities in shoreline and littoral habitats	Observations were recorded from 10 shoreline locations around each lake. Uniform disturbance level criteria used nationwide
Riparian vegetation: reflects ability to buffer lake from influence of upland land use activities	Same as for lake habitat complexity
Lake drawdown exposure: reflects potential loss of littoral habitat and loss of connectivity between littoral and riparian zones due to hydrologic alteration and/or drought	Observations were recorded from 10 shoreline locations around each lake. Information was compared to distribution of drawdown exposure in regional reference sites
Atrazine: provides an indication of exposure to herbicides	Collected from a vertically integrated sample of the upper water column at the open-water site. We report on detection; measured concentrations were compared to an EPA plant-effects benchmark
Chlorophyll *a*: indirect measure of algal biomass, trophic state, and the potential for presence of algal toxins	Collected from a vertically integrated sample of the upper water column at the open-water site. Concentrations were compared to WHO algal toxin benchmark for recreation

Indicator and rationale	Sample location
Methyl mercury: toxic form of mercury that bioaccumulates in the lake food chain	Collected from the top 2 cm of sediment from a core taken from the bottom of the lake. Concentrations were compared to a benchmark
Total mercury: indicates potential exposure and availability of mercury to lake biota	Collected from the top 2 cm of sediment from a core taken from the bottom of the lake. Concentrations were compared to a benchmark
Microcystin: direct measure of algal toxin concentration present on day of sampling	Collected from a vertically integrated sample of the upper water column at the open-water site. We report on detection; measured concentrations were compared to the World Health Organization (WHO) algal toxin benchmark for recreation
Cyanobacteria: includes organisms responsible for release of algal toxins	Collected from a vertically integrated sample of the upper water column at the open-water site. Concentrations were compared to WHO algal toxin benchmark for recreation

Table 1.
Indicators and sampling locations for the national lakes assessment.

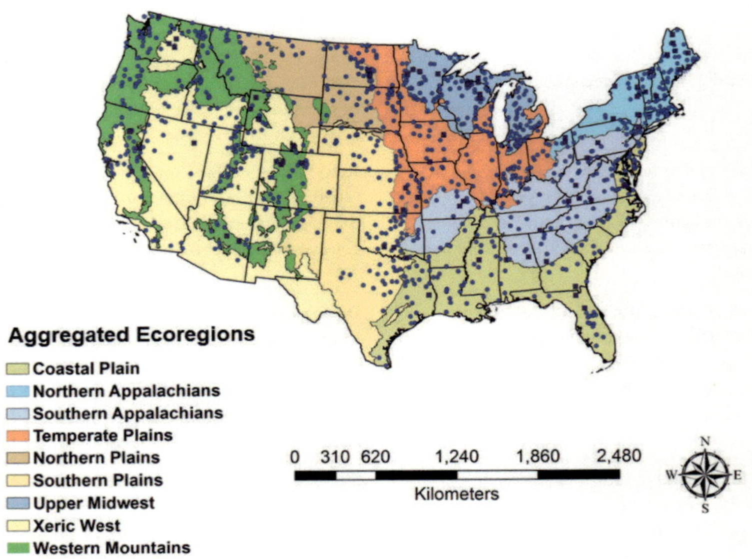

Figure 1.
Distribution of lakes sampled for the 2012 National Lakes Assessment. Circles represent sites selected as part of the probability-based survey design. Squares represent lakes hand selected as additional candidate "least-disturbed" reference sites for use in assigning lake condition categories. Aggregated ecoregions are based on Omernik level 3 ecoregions.

sites sampled to the entire target population of approximately 112,000 lakes targeted by the survey within the conterminous USA. The spatial distribution of sampled lakes in the 2012 survey is shown in **Figure 1**. For more details on survey designs as applied to aquatic resources, see [21, 26–30].

3.3 Response design

The way in which an individual lake is sampled for the various indicators is considered the "response design" [19]. In some cases, as with water samples, this is rather simple. For other indicators, such as physical habitat indicators, the response design is more complex. The NLA consists of two response designs at each lake. A standard single station located at approximately the deepest point in the lake (or midpoint of a reservoir) is used to collect (1) a depth profile of temperature, dissolved oxygen, pH, and conductivity; (2) surface water samples for chemical analyses and phytoplankton; (3) vertical plankton net tows to collect zooplankton; and (4) a sediment core sample. These samples result in data on zooplankton, chlorophyll *a*, acid neutralizing capacity (ANC), conductivity, total nitrogen, total phosphorus, anions/cations, dissolved oxygen, water transparency, temperature, pH, cyanobacteria, atrazine, sediment mercury (total and methyl), and microcystin. Riparian and littoral zone observations are collected at 10 equally spaced locations around the lake perimeter. Benthic macroinvertebrate samples are also collected at these littoral sites around the lake. Details of the collection process can be found in [29] and a similar document tied to each lake survey (**Table 1**).

4. Methods

The methods for the NLA are described in great detail in its supporting documentation (e.g., see [30–34]). A brief summary of critical elements of the approach follows.

4.1 Data acquisition (field and laboratory)

The NLA has developed field protocols intended to be applied consistently at all lakes and reservoirs sampled. This is in contrast to the approach implemented in the European Union to accomplish the objectives of the Water Framework Directive, which employs various methods to arrive at analogous assignments of water body condition (e.g., see [35]). The NLA protocols are also designed to be implemented by field crews who are not all experienced limnologists or aquatic biologists. Many (80–90) field crews (comprised of state and contractor crew employees) are required to sample the selected lakes during a summer sampling window (index period) from June through September. It is important to note that inferences made from the data are estimates of condition found during that index period and do not apply, necessarily, to other parts of the year. In essence, these are "snapshots" of conditions in the lake population during the summer growing season. Standardized field and laboratory protocols are used to collect and process the samples. Standardized field forms, either paper or electronic, are used by the crews to record measurements and observations. The samples that are collected are sent to processing laboratories for analyses. The field and laboratory data are sent to a central repository for inclusion into the data sets (see [30] for details). A comprehensive quality assurance program is developed and implemented for all field, laboratory, data analysis, and data management activities in the NLA to ensure that results are of known and adequate quality to be used in the assessment (e.g., see [33]).

4.2 Indicator development and evaluation

For the benthic macroinvertebrate and zooplankton samples, a comprehensive analysis and evaluation process was used to construct a multimetric index (MMI) of biological integrity for that assemblage. The process was based on general approaches described in [36, 37]. Metrics were developed using autecology information, taxonomic composition, taxonomic diversity, functional feeding groups, habitat preferences and tolerance to disturbance. The rationale and descriptions for each of these indicators can be found in [30, 38–42].

The approach used to measure and describe various dimensions of littoral and riparian physical habitat is described in [43–46]. These measurements result in indicators of lake habitat complexity, shallow water habitat alteration, riparian vegetation cover, lakeshore disturbance, and lake drawdown exposure in the littoral zone [30, 45, 46]. The shallow water habitat alteration indicator is based on visual estimates of the areal cover of several types of natural cover (e.g., snags, macrophytes, overhanging vegetation) observed in the littoral zone around each lake. The riparian vegetation cover indicator is based on visual estimates of vegetation cover and structure in three layers of riparian vegetation observed around each lake. The lakeshore disturbance indicator is based on visual estimates of the presence and proximity of several types of human disturbance (e.g., agricultural activities, residences, marinas) to the lake margin observed around each lake. The lake habitat complexity indicator is based on the mean value of the shallow water habitat alteration and riparian vegetation cover indicators.

For each of the physical, chemical, and biological indicators used in the assessment, a set of benchmarks or thresholds was developed against which to evaluate the quality of the lake relative to that indicator. For the NLA, expected values were developed for each indicator within each of the 9 aggregated ecoregions shown in **Figure 1** based on the distribution of measured values (observed scores), or observed/expected values (calculated scores) of the indicator in the set of least-disturbed reference lakes within that region. Condition thresholds were developed using the 5th and 25th (or 95th and 75th) percentiles of the distribution of the indicator scores in the set of regional reference sites, as described in the NLA 2012 technical report [30], and all sampled sites were assigned to good, fair, or poor condition based on those thresholds. More detailed discussions of the concepts underpinning behind the use of reference sites to model regional or individual lake expected indicator values in least-disturbed reference sites can be found in [25, 45, 47, 48].

4.3 Population estimates

The analytical goal of the assessment is to produce estimates of the number of lakes (or percent of lake number) falling into a condition class or stressor level based on the indicator data and the weights from the survey design [49]. Examples of how this was done for lakes and wetlands are presented in [21, 50]. The weight assigned to an individual lake is an estimate of the number of lakes in the target population represented by that lake and is used to develop a cumulative picture of the total target population. Status of the total lake population can be assessed for each of the indicators measured, whether they are biological, chemical, or physical. These population estimates represent the assessment of biological, chemical, and physical integrity goals expressed in the CWA.

4.4 Ranking of stressors

The final element of the assessment is intended to answer another key NLA question—"What is the relative importance of the different stressors impacting lakes?" This element ranks the potential stressors to biological condition that were measured during the survey. This assessment element is not intended to determine the "cause" of poor conditions at an individual lake but rather to evaluate and then rank the relative improvement in national status that might be gained, biologically, if one were to eliminate the adverse influence of each stressor through policy changes or management efforts. The quantitative approach borrowed from the medical literature to derive relative rankings is outlined in [51, 52]. This approach first requires a "relative extent" estimate (for each stressor) represented by the proportion of lakes in poor condition for that stressor. Then, the "relative risk" to biological indicators associated with poor conditions of each stressor indicators is calculated. Relative risk is the ratio of the percentage of lakes in poor biological condition in the subset of lakes that have high stress (poor condition), divided by the percentage of lakes in poor biological condition in the subset of lakes with stressor condition not classified as poor. Combining relative risk with relative extent of lakes with poor biological condition allows the calculation of "attributable risk," that is, the potential reduction in the percentage of lakes with poor biological condition if all of the lakes with poor stressor condition were to be restored so that they would be in good or fair stressor condition. These estimates are calculated for each stressor indicator and ranked relative to one another to see where the greatest improvement in biological condition might be expected.

5. National and regional status estimates

The results presented here are examples of a few of the ways to present and interpret the results from the NLA. We do not present a comprehensive assessment of lake condition based on NLA results here (see [34]). The first objective of the NLA is to describe the biological integrity of lakes within the conterminous USA. Based on a pelagic zooplankton multimetric index (MMI) of biological integrity, only 53 ± 7% of lakes in the conterminous USA ("National") are considered to be in good condition (**Figure 2**). A greater percentage of the natural lakes are in good condition (61 ± 10%) when compared with man-made lakes (43 ± 8%; **Figure 2**).

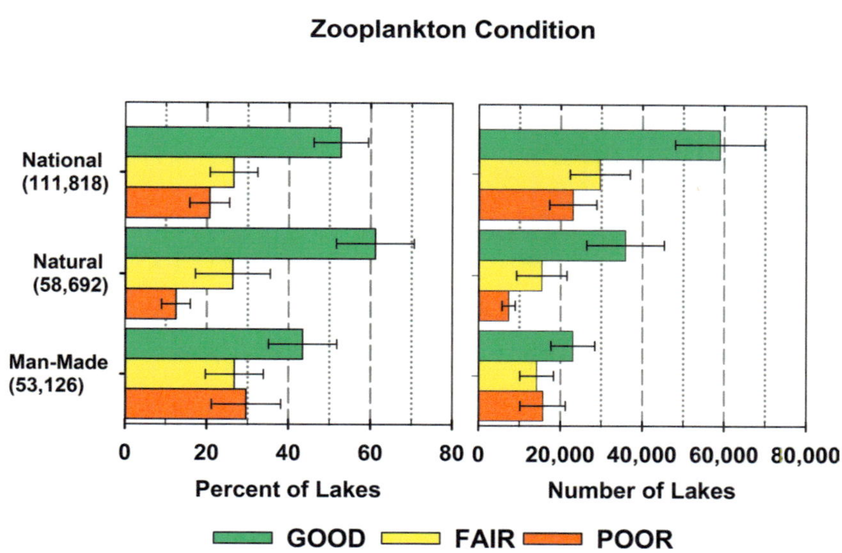

Figure 2.
Status of lake biological condition for the 2012 National Lakes Assessment based on a multimetric index (MMI) for the zooplankton assemblage. Results are presented nationally and by lake origin type (natural versus man-made) in the conterminous United States (i.e., lower 48 states). Estimates are presented as percent of lakes in each condition class (good, fair, or poor relative to regional determination of least-disturbed condition) and as the absolute numbers of lakes. Values in parentheses are the estimated number of target lakes in the population. Error bars are 95% confidence intervals.

Estimates produced for the 9 aggregated ecoregions allow one to consider regional patterns of condition in the context of the national estimates (**Figure 3**). Four regions (the Northern Appalachians, the Upper Midwest, the Southern Plains, and the Western Mountains) have more than 60% of their target population of lakes in good condition based on the zooplankton MMI. Three other regions (the Southern Appalachians, the Northern Plains, and the Xeric West) have a higher percentage of lakes in their target population in poor condition than good condition based on the zooplankton MMI (**Figure 3**).

Comparing regional and national estimates addresses the public's questions about whether poor conditions are distributed uniformly across the country or focused regionally. Such information allows for identifying and prioritizing those areas where the greatest need exists to address a specific problem. However, because the quality of least-disturbed sites varies regionally, direct comparisons among aggregated ecoregions need to be interpreted cautiously in terms of the lake population in one region having "better" (or "worse") lake condition than the lake population in another region.

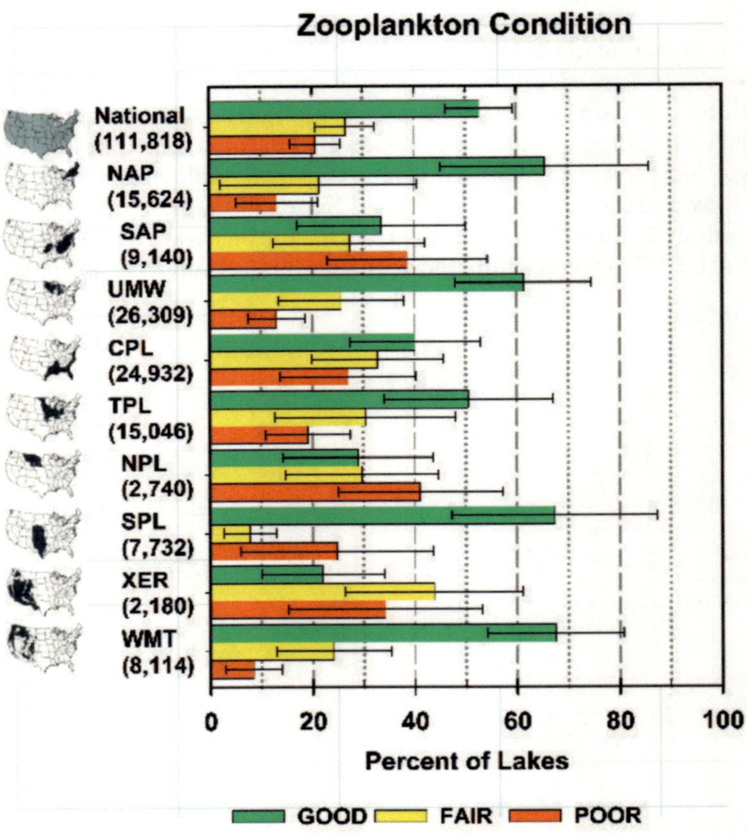

Figure 3.
Status of lake biological condition for the 2012 National Lakes Assessment based on a multimetric index (MMI) for the zooplankton assemblage. Results are presented nationally and for 9 aggregated ecoregions of the conterminous United States (i.e., lower 48 states). Estimates are presented as percent of lakes in each condition class (good, fair, or poor relative to regional determination of least-disturbed condition). Values in parentheses are the estimated number of lakes in the target population. Error bars are 95% confidence intervals. Aggregated ecoregion codes: NAP, Northern Appalachians; SAP, Southern Appalachians; UMW, Upper Midwest; CPL, Coastal Plain; TPL, Temperate Plains; NPL, Northern Plains; SPL, Southern Plains; XER, Xeric West; and WMT, Western Mountains.

Similar assessments can be made for any of the stressor indicators. Lake condition based on two nutrients (total phosphorus and total nitrogen) appears to be similar nationally, with less than 50% of all lakes with nutrient concentrations low enough to be considered in good condition (**Figure 4**). For both nutrients, man-made lakes have a lower percentage of lakes in good condition, and a greater percentage of lakes in poor condition, than natural lakes (**Figure 4**). Despite the fact that regions differ greatly in their proportion of natural versus man-made lakes, the national patterns observed for total phosphorus (TP) and total nitrogen (TN) are remarkably similar for the two types of lakes.

When we focus on a single nutrient, total phosphorus, 45% of the lakes in the conterminous USA are classified in good condition relative to regional expectations (**Figure 5**). Almost 55% of natural lakes were in good condition based on total phosphorus concentrations, compared with about 30% of man-made lakes (**Figure 4**). Across the 9 ecological regions, the Southern Appalachians, the Northern Plains, and Southern Plains exhibited the smallest percentages of lakes in good condition relative to total phosphorus with 23, 10, and 28% of the lakes classified in good condition, respectively (**Figure 5**).

Figure 6 shows comparable results at the national scale for the four measures of physical habitat quality in lakes—lake habitat complexity, shallow water habitat alteration, riparian vegetation cover, and lakeshore disturbance. In each case,

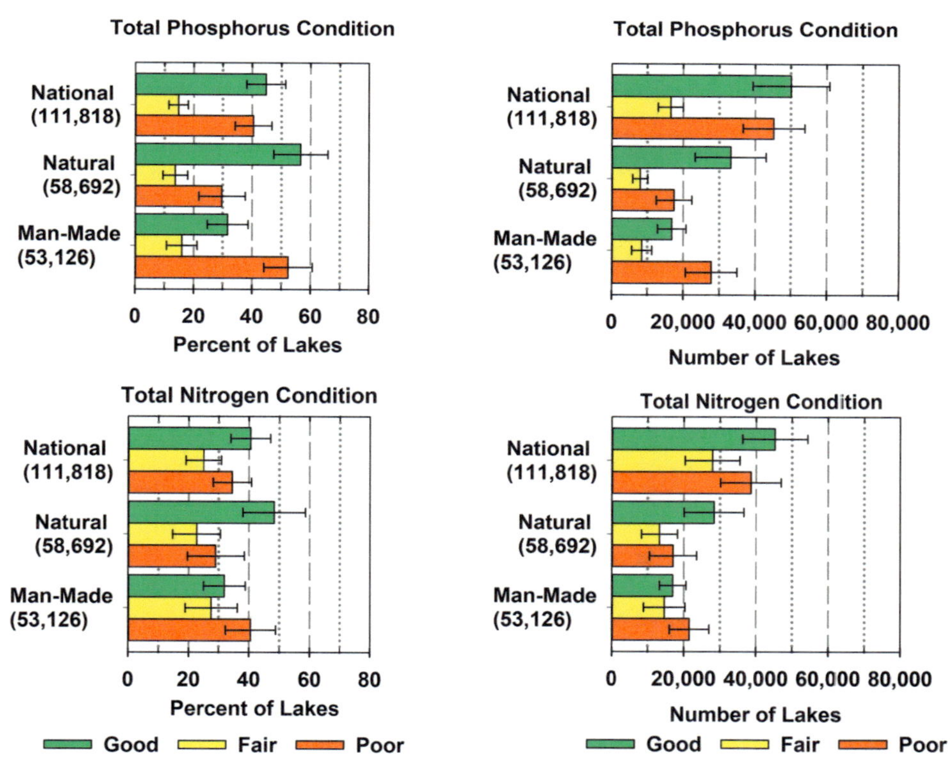

Figure 4.
Status of lake condition for the 2012 National Lakes Assessment based on total phosphorus and total nitrogen concentrations. Results are presented nationally and by lake origin type (natural versus man-made). Estimates are presented as the percent of lakes in each condition class (good, fair, poor relative to regional determination of least-disturbed condition) and as the absolute numbers of lakes. Values in parentheses are the estimated number of target lakes in the population. Error bars are 95% confidence intervals.

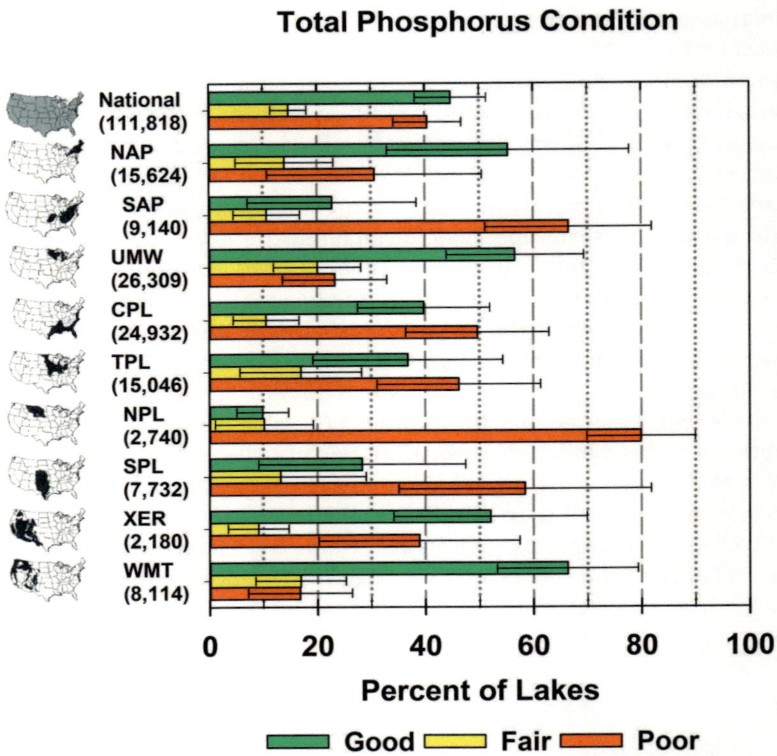

Figure 5.
Status of lake condition for the 2012 National Lakes Assessment based on total phosphorus concentrations. Results are presented nationally and for nine aggregated ecoregions of the conterminous United States (i.e., lower 48). Estimates are presented as percent of lakes in each condition class (good, fair, or poor relative to regional determination of least-disturbed condition). Error bars are 95% confidence intervals. Aggregated ecoregion codes: NAP, Northern Appalachians; SAP, Southern Appalachians; UMW, Upper Midwest; CPL, Coastal Plain; TPL, Temperate Plains; NPL, Northern Plains; SPL, Southern Plains; XER, Xeric West; and WMT, Western Mountains.

no more than 55% of the lakes in the country are in good condition for the respective physical habitat indicator. Nationally, the percent of lakes in good condition ranged from 28% (lakeshore disturbance) to 55% (riparian vegetation cover). Except for the shallow water habitat indicator, the percentage of natural lakes in good condition was greater than the percentage of man-made lakes in good condition.

Lake trophic state is a general indicator of lake productivity; the National Secchi Dip-In [11] provides an excellent overview that is based primarily on [53]. For the NLA, trophic state was estimated using phytoplankton chlorophyll *a* concentration, and condition was assigned using a single set of benchmarks across all ecoregions. **Figure 7** shows that nationally, about 10% of the lakes are classified as oligotrophic (chlorophyll *a* < 2 μg/L), and about 20% of the lakes are classified as hypereutrophic (chlorophyll *a* > 30 μg/L). The population of natural lakes appears to be less productive (i.e., have a larger percentage of lakes classified oligotrophic and mesotrophic) than the population of man-made lakes, which have a greater percentage of lakes classified as eutrophic and hypereutrophic (**Figure 7**). Across the 9 ecoregions, the largest percentage of oligotrophic lakes (nearly 60%) occurs in the Western Mountains (**Figure 8**). The Southern Plains has >40% of lakes classified as hypereutrophic, while the Temperate Plains has >30% of lakes classified as hypereutrophic (**Figure 8**).

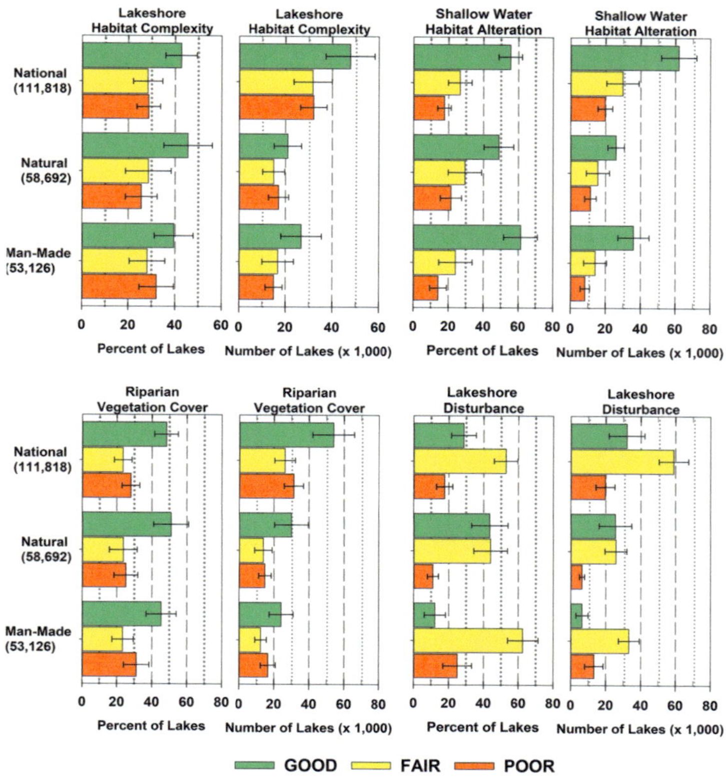

Figure 6.
Status of lake condition based on four indicators of physical habitat measured in the 2012 National Lakes Assessment: lakeshore habitat complexity, shallow water habitat alteration, riparian vegetation cover, and lakeshore disturbance. Results are presented nationally and by lake origin type (natural versus man-made). Estimates are presented as percent of lakes in each condition class (good, fair, poor relative to regional determination of least-disturbed condition) and absolute numbers of lakes. Values in parentheses are the estimated number of lakes in the target population. Error bars are 95% confidence intervals.

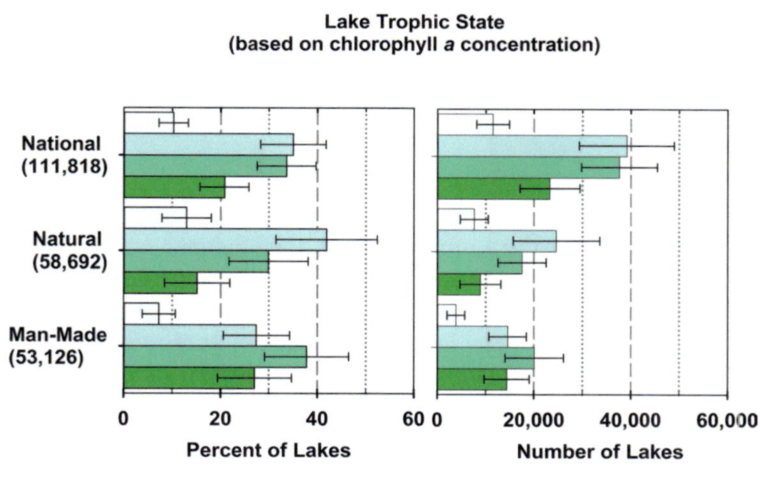

Figure 7.
Status of lake trophic state for the 2012 National Lakes Assessment. Trophic classes are based on chlorophyll a concentration. Results are presented nationally and by lake origin type. Estimates are presented as the percent of lakes and as the absolute number of lakes in each trophic category. Values in parentheses are the estimated number of lakes in the target population. Error bars are 95% confidence intervals.

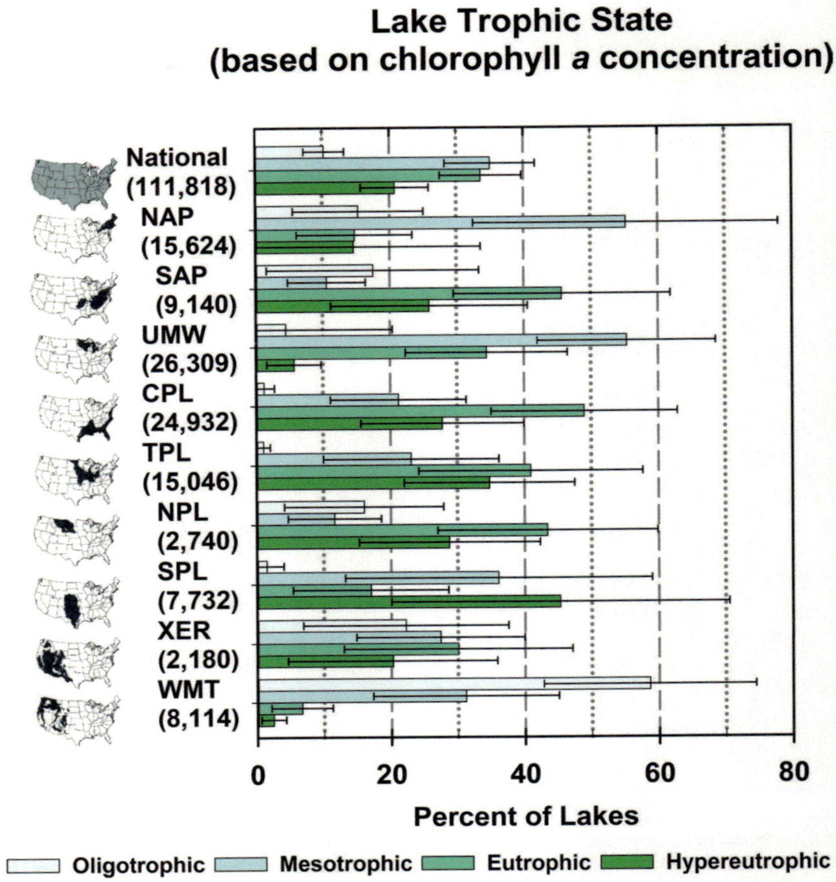

Figure 8.
Status of lake trophic state for the 2012 National Lakes Assessment. Trophic classes are based on chlorophyll a concentration. Results are presented nationally and for nine aggregate ecoregions. Estimates are presented as the percent of lakes in each trophic category. Values in parentheses are the estimated number of lakes in the target population. Error bars are 95% confidence intervals. Aggregated ecoregion codes: NAP, Northern Appalachians; SAP, Southern Appalachians; UMW, Upper Midwest; CPL, Coastal Plain; TPL, Temperate Plains; NPL, Northern Plains; SPL, Southern Plains; XER, Xeric West; and WMT, Western Mountains.

6. Change and trend estimates

Historically, the monitoring community has been focused on tracking trends at individual locations. The historic graphs of CO_2 levels at Mauna Loa [54] and decreases in water clarity resulting from increases in primary productivity in Lake Tahoe [55] are excellent examples. Tracking conditions at individual locations can be quite useful and is akin to tracking the weight or obesity status of an individual (i.e., useful for that individual but their use in large-scale policy discussions depends entirely on the circumstance). The Mauna Loa data clearly provide strong evidence for global increases in CO_2 given atmospheric circulation. In contrast, the isolated nature of individual lakes such as Lake Tahoe does not lend support for interpreting the Lake Tahoe water clarity data as a signal of a national or a global increase in lake productivity. The changes and trends that the NLA seeks to track are population trends...conceptually similar to asking the human health question

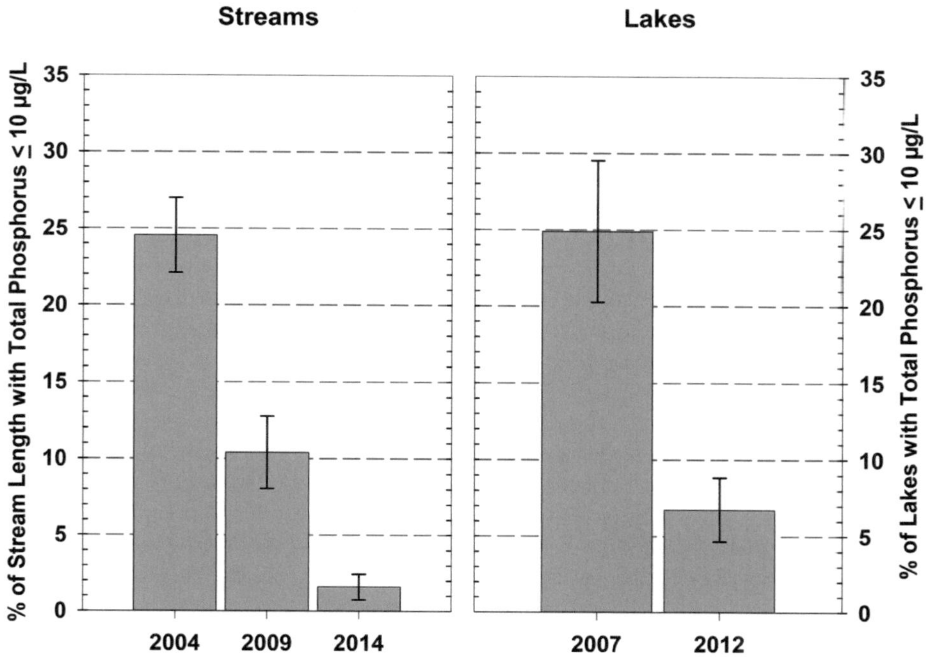

Figure 9.
Changes in total phosphorus (TP) in dilute streams and lakes across the conterminous USA based on the initial surveys of the National Rivers and Streams Assessment and the National Lakes Assessment. Data from [55]. Error bars are 95% confidence intervals.

"Has the number or percent of the population that is obese increased?". For the NLA, that translates to "Has the percent of lakes in poor (or good) condition class changed over time?"...essentially, is there a change in status over time? The current NLA online tools and reports show both status and changes. The best published examples of the intent of the NLA are [56, 57]. In [57], the authors document an increase in total phosphorus across the country that is especially evident in low nutrient lakes and streams. **Figure 9** displays the results discussed in that paper, showing that over the three initial stream surveys conducted as part of the NARS, the percentage of the total length of the stream population that had total phosphorus concentrations less than 10 μg/L decreased from 24.5% to just 1.6% between 2004 and 2014. Lakes were only surveyed twice during this period and showed a similar pattern with 24.9% of lakes with total phosphorus concentrations below 10 μg/L in 2007 decreasing to 6.7% of the lakes in that low nutrient category in 2012 (**Figure 9**). While it may be too early to know if these unidirectional changes and trends will persist, they are excellent examples of the types of population changes and trends that the NLA (and the NARS assessments in general) are intended to identify.

7. Stressor rankings

While the results presented above are useful for describing status and trends in lake conditions, they do not address the potential associations of different stressors with biological condition. When studying individual lakes, we are used to asking questions about the cause or combination of causes of the problem we have found. This is similar to asking "Why am I over-weight or gaining weight?" In population-level or policy-level discussions, it is not about finding a specific cause of problems, but rather finding some way to rank the various causes. In the context of assessing obesity, of all the causes of increasing weight in the U.S., what is their relative importance, and which would result in the largest improvements in the obesity situation if it were tackled? The NLA, and the NARS more broadly, have adapted tools from the human health field (relative risk and attributable risk) to address this question [51, 52].

Three pieces of information are needed to rank stressors according to importance and pervasiveness. The first is relative extent—a measure of how widespread a particular stressor or potential cause of problems is. How many lakes, for example, have high (or poor) levels of total phosphorus? This is shown in the left panel of **Figure 10**. From the figure, one can see that 40% of the lakes have total phosphorus at levels high enough to be considered poor. Similar information is presented for the other stressors nationally and separately for natural and man-made lakes.

The second piece of information is an estimate of the relative risk posed to biological condition (e.g., as assessed using the zooplankton MMI) by each stressor (**Figure 10**, center panel). This provides an estimate of the impact of a particular stressor on the zooplankton community when the stressor occurs at high levels (poor stressor condition). At a relative risk of 1, zooplankton are equally likely to be in poor condition if the stressor is at high levels (poor stressor condition) or at low to medium levels (good and fair stressor condition). At a relative risk of 2,

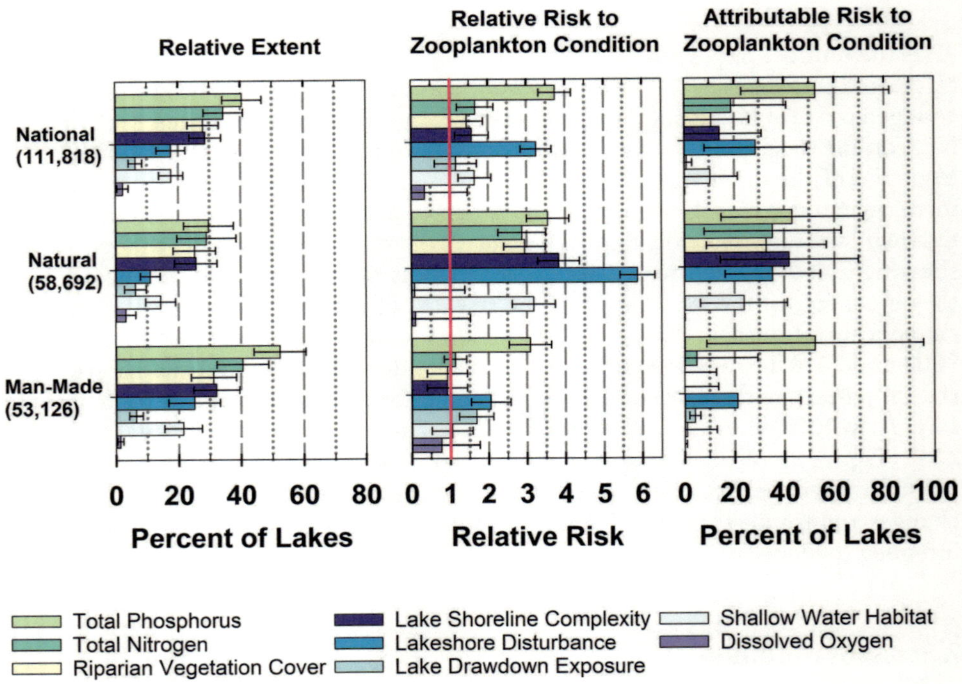

Figure 10.

Estimates for ranking stressors relative to their impact on the zooplankton assemblage for the 2012 National Lakes Assessment. Results are presented nationally and by lake origin type. Solid line represents a relative risk of 1, below which a stressor poses no risk to the biological assemblage. Error bars are 95% confidence intervals.

the zooplankton community is two times as likely to be in poor condition in the presence of high stressor levels as it is to be in poor condition with low to medium levels of the stressor. Nationally, zooplankton communities are more than 3.5 times as likely to be in poor condition with high levels of total phosphorus than with low to medium levels of total phosphorus.

The third piece of information combines the relative extent values and relative risk values to generate an attributable risk (AR) estimate (**Figure 10**, right panel). This answers the question: "How much of an improvement in lake biological condition would be seen if all the total phosphorus values in poor condition were improved to fair or good condition?". In the case of the potential risk of total phosphorus to lake biological condition as represented by the zooplankton community, we would expect a 52% reduction in the number of lakes in the target population in poor biological condition for zooplankton if the total phosphorus concentrations in these lakes were decreased enough to change the stressor condition from poor to either fair or good. The point of calculating the attributable risk is to generate an estimate of the potential benefit in zooplankton communities determined the same way for all stressors. Ranking via AR allows a consistent and relevant approach for providing a relative ranking of the stressors. **Figure 10** suggests that for natural and man-made lakes combined, the greatest potential benefit to the pelagic zooplankton community would result from nutrient control or reducing lakeshore disturbance. In natural lakes, the attributable risks to zooplankton from poor shoreline habitat complexity, riparian vegetation condition, excessive shoreline disturbance, and nutrients (total nitrogen and total phosphorus) are all at high values (between 32 and 43%). These results are consistent with abundant research showing that near-shore habitat alteration and increased nutrient loading are associated, and further suggest that near-shore habitat protection and restoration may be a fruitful strategy for controlling nutrients and improving zooplankton biointegrity.

8. Tracking specific threats and emerging threats

Among the biggest challenges and frustrations in monitoring is the time lag in addressing specific or new threats. When the acid rain issues arose in the 1980s, among the first questions raised was "How big is the problem?". Sadly, a reluctance to invest the time to assemble the technical experts to design and then implement a survey prompts premature policy decisions in the absence of solid information. While it is not possible to design a survey that anticipates every single problem that will arise, it is possible to design a survey that answers key questions about the health of our lakes and the relative importance of currently known stressors. The NLA does this well, in part because of the flexibility to adapt the survey design to new threats (e.g., see [58]). Additionally, the NLA serves as a platform from which to launch initial investigations into emerging issues to understand the nature of their size and distribution as well as track past and ongoing threats. The NLA continues to track the trophic state of lakes across the country (**Figure 10**). While the specific cause of eutrophication may have shifted from point sources to nonpoint sources it is still important to track this status as a key measure of how we manage our lakes. As other threats emerge, the NLA provides a platform to track their extent in lakes. Currently, harmful algal blooms and the toxins they produce (e.g., microcystin), mercury, and atrazine are among the specific stressors being tracked via NLA. The NLA 2012 website [59] has excellent presentations to explore the breadth of these threats.

9. Conclusions

Until the early 2000s, the history of monitoring lakes in the United States had been a succession of reactive efforts to assess particular stressors to determine how widespread they were and what policies, if any, should be adopted to tackle them. This strategy was moderately effective with domestic point source discharges like sewage treatment plants and with the deposition of acidic compounds as a result of sulfur and nitrogen compounds into the atmosphere. But many stressor-response problems are more complex, both in regional distribution and in likely causes. The NLA was initiated in 2007 to provide a more holistic and comprehensive approach to monitoring the quality of our lakes and the stressors impacting them while still allowing a platform to track specific lake stressors of concern as they emerge.

Acknowledgements

The authors gratefully acknowledge the tremendous effort and collaboration on the part of the USEPA Office of Water, the USEPA Office of Research and Development, and numerous state, federal, tribal, and contractor organizations that provide field, laboratory, and data analysis support for the National Lakes Assessment. The authors appreciate the constructive reviews of earlier drafts by D. Adams, F. John, and J. Markwiese. This chapter has been subjected to Agency review and has been approved for publication. The views expressed in this chapter are those of the author(s) and do not necessarily reflect the views or policies of the U.S. Environmental Protection Agency. Any mention of trade names, products, or services does not imply an endorsement by the U.S. Government or the U.S. Environmental Protection Agency. The EPA does not endorse any commercial products, services, or enterprises.

Conflict of interest

The authors declare no conflict of interest.

Author details

David V. Peck[1*], Steven G. Paulsen[1], Philip R. Kaufmann[1,2] and Alan T. Herlihy[2,3]

1 Office of Research and Development, US Environmental Protection Agency, Corvallis, OR, USA

2 Department of Fisheries and Wildlife, Oregon State University, Corvallis, OR, USA

3 National Research Council Senior Fellow, Corvallis, OR, USA

*Address all correspondence to: peck.david@epa.gov

References

[1] MacGonigle JN. The geography of the Southern Peninsula of the United States. National Geographic. 1896;7(12):381-394. Available from: https://archive.org/details/nationalgeograph71896nati/page/382/mode/2up/search/Lakes+of+Florida [Accessed: 03 March 2020]

[2] Federal Water Pollution Control Act—Amendments of 1972. Pub. Law No. 33 U.S.C. §1251 et seq. 1972. Available from: https://www.govinfo.gov/content/pkg/STATUTE-86/pdf/STATUTE-86-Pg816.pdf [Accessed: 24 March 2020]

[3] H. John Heinz Center for Science, Economics, the Environment. The state of the Nation's Ecosystems: Measuring the Lands, Waters, and Living Resources of the United States. New York: Cambridge University Press; 2002. 288 p

[4] O'Malley R, Negra C. Filling the Gaps: Priority Data Needs and Key Management Challenges for National Reporting on Ecosystem Condition. Washington, DC: The H. John Heinz Center for Science, Economics and the Environment; 2006. 111 p. Available from: https://www.researchgate.net/publication/307509444_Filling_the_gaps_Priority_data_needs_and_key_management_challenges_for_national_reporting_on_ecosystem_condition [Accessed: 24 March 2020]

[5] National Research Council. Environmental Monitoring. Vol. IV. Washington, DC: National Academy of Sciences; 1977

[6] Paulsen SG, Hughes RM, Larsen DP. Critical elements in describing and understanding our nation's aquatic resources. Journal of the American Water Resources Association. 1998;**34**(5):995-1005

[7] GAO (Government Accountability Office). Water Quality: Key EPA and State Decisions Limited by Inconsistent and Incomplete Data. Gaithersburg, Maryland: Resources, Community, and Economic Development Division, Government Accountability Office; 2000. Report No.: GAO/RCED-00-54

[8] GAO (Government Accountability Office). Better Monitoring Techniques are Needed to Assess the Quality of Rivers and Streams. Vol. 1. Washington, DC: Community, and Economic Development Division, Government Accountability Office; 1981. Report No.: GAO/CED-81-30

[9] National Research Council. Ecological Indicators for the Nation. Washington, DC: National Academy Press; 2000

[10] Gakstatter JH, Allum MO, Omernik JM. Lake Eutrophication: Results from the National Eutrophication Survey. Corvallis, Oregon: U.S. Environmental Protection Agency, Office of Research and Development, Laboratory CER; 1975. Report No.: EPA 645/R-75/001

[11] North American Lake Management Society. National Secchi Dip-In webpage: NALMS. 2019. Available from: https://www.nalms.org/secchidipin/ [Accessed: 20 February 2020]

[12] Landers DH, Overton WS, Linthurst RA, Brakke DF. Eastern Lake survey, regional estimates of lake chemistry. Environmental Science & Technology. 1988;**22**(2):128-135

[13] Eilers JM, Brakke DF, Landers DH, Kellar PE. Characteristics of lakes in mountainous areas of the western United States. Verhandlungen der Internationalen Vereinigung fur Theoretische und Angewandte Limnologie. 1988;**23**:144-151

[14] Kaufmann PR, Herlihy AT, Mitch ME, Messer JJ, Overton WS.

Stream chemistry in the eastern United States: Synoptic survey design, acid-base status, and regional patterns. Water Resources Research. 1991;**27**:611-627

[15] Hughes RM, Paulsen SG, Stoddard JL. EMAP-surface waters: A multiassemblage, probability survey of ecological integrity in the U.S.A. Hydrobiologia. 2000;**422/423**:429-443

[16] Larsen DP, Stevens DL, Selle AR, Paulsen SG. Environmental monitoring and assessment program: EMAP-surface waters, a northeast lakes pilot. Lake and Reservoir Management. 1991;**7**(1):1-11

[17] Whittier TR, Larsen DP, Peterson SA, Kincaid TM. A comparison of impoundments and natural drainage lakes in the Northeastern USA. Hydrobiologia. 2002;**470**:157-171

[18] Whittier TR, Paulsen SG, Larsen DP, Peterson SA, Herlihy AT, Kaufmann PR. Indicators of ecological stress and their extent in the population of northeastern lakes: A regional-scale assessment. Bioscience. 2002;**52**(3):235-247

[19] Hughes RM, Peck DV. Acquiring data for large aquatic resource surveys: The art of compromise among science, logistics, and reality. Journal of the North American Benthological Society. 2008;**27**(4):837-859

[20] Peterson SA, Urquhart NS, Welch EB. Sample representativeness: A must for reliable regional lake condition estimates. Environmental Science & Technology. 1999;**33**(10):1559-1565

[21] Peck DV, Olsen AR, Weber MH, Paulsen SG, Peterson C, Holdsworth SM. Survey design and extent estimates for the National Lakes Assessment. Freshwater Science. 2013;**32**(4):1231-1245

[22] Downing JA. Emerging global role of small lakes and ponds: Little things mean a lot. Limnetica. 2010;**29**(1):9-24

[23] Downing JA, Prairie YT, Cole JJ, Duarte CM, Tranvik LJ, Striegl RG, et al. The global abundance and size distribution of lakes, ponds, and impoundments. Limnology and Oceanography. 2006;**51**(5):2388-2397

[24] USEPA (United States Environmental Protection Agency). Visiting a Regional Office Webpage. Washington, DC: US EPA; 2020. Available from: https://www.epa.gov/aboutepa/visiting-regional-office [Accessed: 13 March 2020]

[25] Herlihy AT, Paulsen SG, Sickle JV, Stoddard JL, Hawkins CP, Yuan LL. Striving for consistency in a national assessment: The challenges of applying a reference-condition approach at a continental scale. Journal of the North American Benthological Society. 2008;**27**(4):860-877

[26] Larsen DP. Sample survey design issues for bioassessment of inland aquatic ecosystems. Human and Ecological Risk Assessment: An International Journal. 1997;**3**(6):979-991

[27] Stevens DL, Olsen AR. Spatially restricted surveys over time for aquatic resources. Journal of Agricultural, Biological and Environmental Statistics. 1999;**4**:415-428

[28] Stevens DL Jr, Olsen AR. Variance estimation for spatially balanced samples of environmental resources. Environmetrics. 2003;**14**:593-610

[29] Stevens DL Jr, Olsen AR. Spatially balanced sampling of natural resources. Journal of the American Statistical Association. 2004;**99**(465):262-278

[30] USEPA (United States Environmental Protection Agency). National Lakes Assessment 2012:

Technical Report. Washington, DC: U.S. Environmental Protection Agency, Office of Water; 2017. Report No.: EPA/841/R-16/114

[31] USEPA (United States Environmental Protection Agency). 2012 National Lakes Assessment Field Operations Manual. Washington, DC: U.S. Environmental Protection Agency, Office of Water; 2012. Report No.: EPA/841/B-11/003

[32] USEPA (United States Environmental Protection Agency). 2012 National Lakes Assessment Field Operations Manual. Washington, DC: U.S. Environmental Protection Agency, Office of Water; 2012. Report No.: EPA/841/B-11/004

[33] USEPA (United States Environmental Protection Agency). 2012 National Lakes Assessment. Quality Assurance Project Plan. Washington, DC: U.S. Environmental Protection Agency, Office of Water. 2012. Report No.: EPA 841-B-11-006

[34] USEPA (United States Environmental Protection Agency). National Lakes Assessment 2012: A Collaborative Survey of Lakes in the United States. Washington, DC: U.S. Environmental Protection Agency, Office of Water and Office of Research and Development; 2016. Report No.: EPA 841/R-16/113

[35] Birk S, Bonne W, Borja A, Brucet S, Courrat A, Poikane S, et al. Three hundred ways to assess Europe's surface waters: An almost complete overview of biological methods to implement the Water Framework Directive. Ecological Indicators. 2012;**18**:31-41

[36] Stoddard JL, Herlihy AT, Peck DV, Hughes RM, Whittier TR, Tarquinio E. A process for creating multimetric indices for large-scale aquatic surveys. Journal of the North American Benthological Society. 2008;**27**(4):878-891

[37] Van Sickle J. Correlated metrics yield multimetric indices with inferior performance. Transactions of the American Fisheries Society. 2010;**139**:1802-1917

[38] Hughes RM, Noss RF. Biological diversity and biological integrity: Current concerns for lakes and streams. Fisheries. 1992;**17**(3):11-19

[39] O'Connor RJ, Walls TE, Hughes RM. Using multiple taxonomic groups to index the ecological condition of lakes. Environmental Monitoring and Assessment. 2000;**61**(2):207-229

[40] Stemberger RS, Larsen DP, Kincaid TM. Sensitivity of zooplankton for regional lake monitoring. Canadian Journal of Fisheries and Aquatic Sciences. 2001;**58**:2222-2232

[41] Stemberger RS, Lazorchak JM. Zooplankton assemblage responses to disturbance gradients. Canadian Journal of Fisheries and Aquatic Sciences. 1994;**51**:2435-2447

[42] Stemberger RS, Miller EK. A zooplankton-N:P-ratio indicator for lakes. Environmental Monitoring and Assessment. 1998;**51**(1):29-51

[43] Kaufmann PR, Hughes RM, Van Sickle J, Whittier TR, Seeliger CW, Paulsen SG. Lakeshore and littoral physical habitat structure: A field survey method and its precision. Lake and Reservoir Management. 2014;**30**(2):157-176

[44] Kaufmann PR, Hughes RM, Whittier TR, Bryce SA, Paulsen SG. Relevance of lake physical habitat indices to fish and riparian birds. Lake and Reservoir Management. 2014;**30**(2):177-191

[45] Kaufmann PR, Peck DV, Paulsen SG, Seeliger CW, Hughes RM, Whittier TR, et al. Lakeshore and littoral physical habitat structure in a national lakes

assessment. Lake and Reservoir Management. 2014;**30**(2):192-215

[46] Fergus CE, Brooks JR, Kaufmann PR, Herlihy AT, Pollard AI, Weber MH, et al. Lake water levels and associated hydrologic characteristics in the conterminous U.S. Journal of the American Water Resources Association. 2020. DOI: 10.1111/1752-1688.12817

[47] Herlihy AT, Sobota JB, McDonnell TC, Sullivan TJ, Lehmann S, Tarquinio E. An a priori process for selecting candidate reference lakes for a national survey. Freshwater Science. 2013;**32**(2):385-396

[48] Stoddard JL, Larsen DP, Hawkins CP, Johnson RK, Norris RH. Setting expectations for the ecological condition of streams: The concept of reference condition. Ecological Applications. 2006;**16**:1267-1276

[49] Kincaid TM, Olsen AR. Spsurvey: Spatial survey design and analysis. R Package version 3.3. 2016. Available from: https://cran.r-project.org/web/packages/spsurvey/index.html [Accessed: 24 March 2020]

[50] Olsen AR, Kincaid TM, Kentula ME, Weber MH. Survey design to assess condition of wetlands in the United States. Environmental Monitoring and Assessment. 2019;**191**(1):268

[51] Van Sickle J, Stoddard JL, Paulsen SG, Olsen AR. Using relative risk to compare the effects of aquatic stressors at a regional scale. Environmental Management. 2006;**38**(6):1020-1030

[52] Van Sickle J, Paulsen SG. Assessing the attributable risks, relative risks, and regional extents of aquatic stressors. Journal of the North American Benthological Society. 2008;**27**(4):920-931

[53] Carlson RE, Simpson J. A Coordinator's Guide to Volunteer Lake Monitoring Methods. Madison, Wisconsin: North American Lake Management Association; 1996

[54] Earth Systems Research Laboratory GMD. Global Greenhouse Gas Reference Network-Trends in Atmospheric Carbon Dioxide. Boulder, Colorado: NOAA; 2020. Available from: https://www.esrl.noaa.gov/gmd/ccgg/trends/weekly.html [Accessed: 18 March 2020]

[55] Tahoe Environmental Research Center. Tahoe: State of the Lake Report 2019. Davis, California: University of California-Davis; 2019. Available from: https://tahoe.ucdavis.edu/stateofthelake [Accessed: 19 March 2020]

[56] Larsen DP, Urquhart NS, Kugler DL. Regional-scale trend monitoring of indicators of trophic condition of lakes. Water Resources Bulletin. 1995;**31**:117-139

[57] Stoddard JL, Van Sickle J, Herlihy AT, Brahney J, Paulsen S, Peck DV, et al. Continental-scale increase in lake and stream phosphorus: Are oligotrophic systems disappearing in the United States? Environmental Science & Technology. 2016;**50**(7):3409-3415

[58] Ferrey ML, Heiskary S, Grace R, Hamilton MC, Lueck A. Pharmaceuticals and other anthropogenic tracers in surface water: A randomized survey of 50 Minnesota lakes. Environmental Toxicology and Chemistry. 2015;**34**(11):2475-2488

[59] USEPA. National Lakes Assessment 2012 webpage. Washington, DC: U.S. Environmental Protection Agency; 2016. Available from: https://www.epa.gov/national-aquatic-resource-surveys/national-lakes-assessment-2012-results [Accessed: 20 February 2020]

Chapter 7

Rivers and Streams: Upgrading Monitoring of the Nation's Freshwater Resources - Meeting the Spirit of the Clean Water Act

Steven G. Paulsen, David V. Peck, Philip R. Kaufmann and Alan T. Herlihy

Abstract

The goal of the Clean Water Act (CWA) is to restore and maintain the chemical, physical and biological integrity of the waters in the United States. Much of the monitoring and assessment is reasonably delegated to the States to monitor and report the condition of their water to Congress through the Environmental Protection Agency. States have historically been fully occupied in monitoring the most egregious water quality problems along with select high priority water bodies. This approach, while addressing State priorities with finite resources, does not capture the full spectrum and scope of water quality conditions within and across State boundaries. Hence, the reporting on progress in meeting the goals of the CWA has not been realized. In this chapter, we describe the partnership between EPA, the States and Tribes to remedy this information gap for rivers and streams. Filling this gap requires both improved monitoring designs to reflect conditions across all waters as well as the expansion of indicators to move beyond water chemistry to include all three elements of the CWA goal—chemical, physical and biological integrity.

Keywords: streams, rivers, monitoring, assessment, National Rivers and Streams Assessment, United States, ecological indicators, survey design, National Aquatic Resource Assessments, water quality, biological integrity, physical habitat, Clean Water Act

1. Introduction

Access to credible, quantitative information regarding the status and trends in water resource conditions is essential for the development of effective national policies for managing water resources in the United States. The US Clean Water Act (CWA) expresses the national desire to restore and maintain the chemical, physical, and biological integrity of US waters and requires that information on status and trends be reported [1]. The need and desire to improve the quality of water resource assessments is not peculiar to the US. Australia has made assessment and management of its aquatic resources a major national focus [2–4]. The Water Framework Directive instituted by the European Community includes key components that are a general requirement for ecological protection and a general minimum chemical standard that is applicable to all surface waters [5]. An assessment of major river basins by 2007 was also called for in the Water Framework Directive [6]. Dwindling budgets for environmental protection, particularly for monitoring and assessment, suggest that all countries will face both technical and fiscal challenges of how to provide assessments that quantify water resource conditions over continental scales. Similar approaches to incorporating chemical, physical and biological information into assessments of individual (e.g., a single river reach) have been adopted by many countries. Much of the technical work in the US and elsewhere has focused on developing biological indicators (e.g., [7–11]). However, it remains unclear if improvements in the science of monitoring survey design have been adopted or implemented. In the US, randomized sampling designs are considered a critical element in support of regional and national surveys (e.g., [12, 13]) because the use of such designs provides a rigorous inference protocol for extending assessments of individual sites to the entire population of the water resource of interest.

The passage of the Clean Water Act (CWA) amendments to protect US water resources in 1972 [14] was an historic event resulting in a law that served as the gold standard for environmental protection globally. Two sections of the CWA stand out with respect to monitoring and assessment. Section 303(d) calls for States to develop a list of waterbodies that fail to support their designated use and to conduct a "Total Maximum Daily Load" (TMDL) analysis for these waterbodies...a total maximum daily load below which the offending "pollutant" should be kept in order to restore designated use. Under Section 305(b), States report to the US Environmental Protection Agency (EPA), which then reports to Congress and the public on the condition of the States' waters, the success or failure, if you will, of efforts to protect and restore waters. In spite of these reporting efforts, reviews of water quality monitoring programs in the US over the years have concluded that neither EPA nor any other U.S. federal agency was able to provide Congress and the public with an adequate assessment regarding the condition of US water bodies [1, 15–22]. These reviews pointed to a host of factors contributing to the problem. Chief among them were the lack of standardization in monitoring approaches, designs, field and laboratory protocols, and indicators used for assessments. To bridge this information gap, the EPA, States, and Tribes, began collaborating on a monitoring effort to produce assessments that provide the public with improved water-quality information at the national and regional scales - the National Aquatic Resource Surveys (NARS). The NARS includes surveys and assessments describing four major water resource types: estuaries, lakes and reservoirs, wetlands, and rivers and streams. This chapter describes one component of the NARS, the National Rivers and Streams Assessment (NRSA), discussing the origins, evolution and initial results.

The NRSA began as a concept in 2002. The EPA Office of Water (OW) wanted to produce a national assessment for one waterbody type. The funds were insufficient to conduct a full national survey. EPA's Office of Research and Development (ORD) had been partnering with the EPA Regional Offices and States in the western half of the US to evaluate approaches to monitoring and assessing rivers and streams across broad geographic scales [23]. A decision was made to use the data collected on wadeable streams in the western pilot study and combine them with a new effort to collect data on wadeable streams in the eastern half of the country using the same survey design, field and laboratory methods, and assessment approach. This collaboration resulted in the Wadeable Streams Assessment (WSA), the first nationally consistent, statistically rigorous study of US wadeable streams [24, 25]. The EPA and its State partners published the approach and findings of the WSA in a special issue of the Journal of the North American Benthological Society (JNABS, 2008,

Issue 27 now named Freshwater Science). Following the WSA, the EPA and the State partners expanded beyond "wadeable streams" to include all flowing waters in the National Rivers and Streams Assessments (NRSA). The first NRSA survey was conducted in 2008–2009 and has repeated every 5 years thereafter (2013–2014 and 2018–2019 at the time of this writing). This chapter uses the results from the 2013–2014 NRSA survey. We describe insights into the conceptual approach and methods used to make NRSA the only monitoring effort to fulfill the original promise of the CWA for reporting on our success or failure in restoring and maintaining the physical, chemical and biological integrity of the nation's rivers and streams.

2. Methods

2.1 Study area

The focus of NRSA 2013–2014 survey is perennial rivers and streams of the 48 conterminous states. While Alaska and Hawaii are not included in NRSA yet, pilot studies have been conducted in both States and will, hopefully, lead to inclusion of these two states in future assessments [26]. This area covers 7,788,958 km^2 and includes rivers and streams running through private, state, tribal, and federal land.

2.2 Survey design

Sampling locations were selected for the NRSA with a state-of-the-art sample survey design approach [12, 26]. Statistically designed sample surveys have been used in a variety of fields (e.g., election polls, forest inventory analysis, national wetlands inventory) to determine the status of resources of interest (e.g., voter preferences, timber availability, and wetland acreage). Sample surveys have been a tool of choice in a variety of fields when it's essential to be able to make unbiased estimates of the characteristics of a large population by sampling a representative set of a relatively small percentage of sites. Because randomization is incorporated into the sample site selection, the estimates are accompanied by robust estimates of the uncertainty. This approach is especially cost-effective when the population is so large that not all components can be sampled. The target population for the NRSA was the perennial rivers and streams in the conterminous US. To identify the location of all perennial streams, the NRSA design team used the National Hydrography Dataset (NHD-Plus; [27]), a comprehensive set of digital spatial data on surface waters at the 1:100,000 scale For 2008–2009, the NRSA findings represent roughly 1.2 million miles or 1.9 million kilometers of perennial rivers and streams [28].

For each NRSA survey, approximately 1800 sites to be sampled are allocated based on the density of river and stream length across the aggregated ecoregions and States (**Figure 1**), and 10 EPA regions [29]. The intent of the design is to provide more sampling in areas of high river and stream length and less sampling in areas with less length of flowing water. The entire design process (i.e., site selection and weighting during analyses) enables unbiased assessment results (including estimates of uncertainty) that are representative of the condition of the streams and rivers throughout the region and the nation.

For the NRSA, results are reported at three scales: national, three major landform and climatic reporting regions (**Figure 2A**), and nine ecological regions (aggregations of Omernik Level III ecoregions; **Figure 2B**). While not frequently used for reporting in the periodic assessments, the NRSA has sufficient sample sizes to assess condition in each of the 10 EPA regions [29] and in at least 12 of the 18 major hydrologic basins across the conterminous US. For this chapter, results

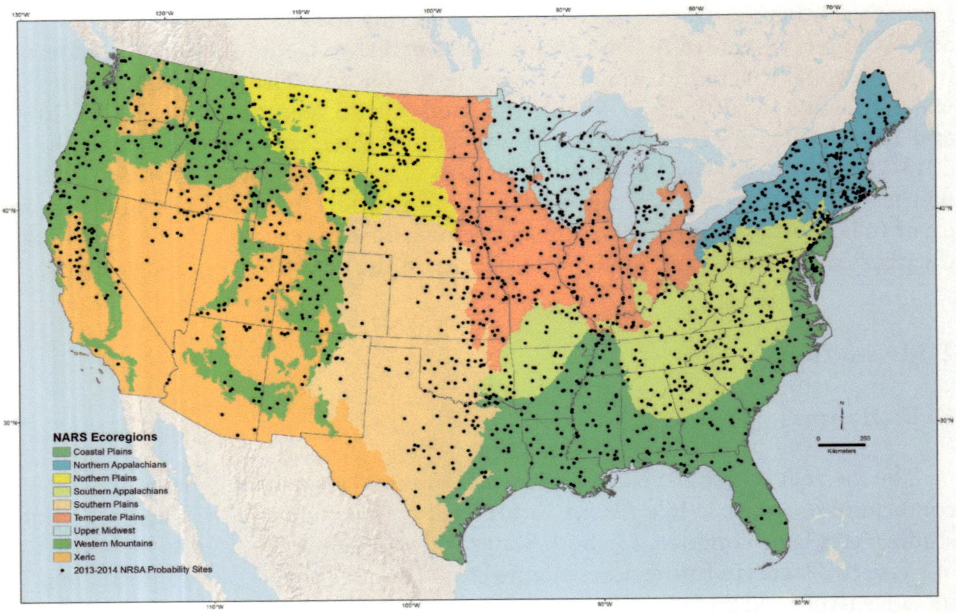

Figure 1.
Locations of the 1853 randomly selected sites sampled in the 2013–2014 National Rivers and Streams Assessment. NARS = National Aquatic Resource Surveys.

for the conterminous U.S. and the three climatic regions are presented as examples of assessment outputs that the NRSA produces. For more detailed results at finer spatial scales see [30].

2.3 Field sampling

Each site is sampled by a 2- to 4-person field crew during a low-flow index period (typically summer) [31]. More than 80 trained crews sampled 1853 random stream and river sites with standardized field protocols over the course of the 2013–2014 field seasons. The field protocols are designed to produce comparable data regarding the ecological condition of stream and river resources and the key stressors at all sites [32, 33].

During each site visit, crews use standardized field procedures to lay out the sample reach and systematically spaced transects to guide data collection [32]. For stream and river sites that require a boat, crews follow a conceptually similar process but are limited to one pass sampling in a downstream direction [33]. Crews record site data and instream and riparian physical habitat measurements on standardized field forms or electronic field recorders for each site. In addition to comprehensive pre-field season training, the proficiency of each crew is evaluated early in the field season, and 10% of the sites are revisited as part of the quality assurance plan for the survey [34].

Field crews collect information in two categories. The first category includes samples that require shipping to a laboratory for additional processing. This includes water samples for chemical and "chemical-like" data (e.g., algal pigments), and for biological samples (i.e., fish, benthic macroinvertebrates and periphyton). The second category includes data that are recorded in the field on standardized electronic forms. The physical habitat data originate as measurements and observations made in the field. These are then forwarded to staff scientists that process the data into metrics and indicators.

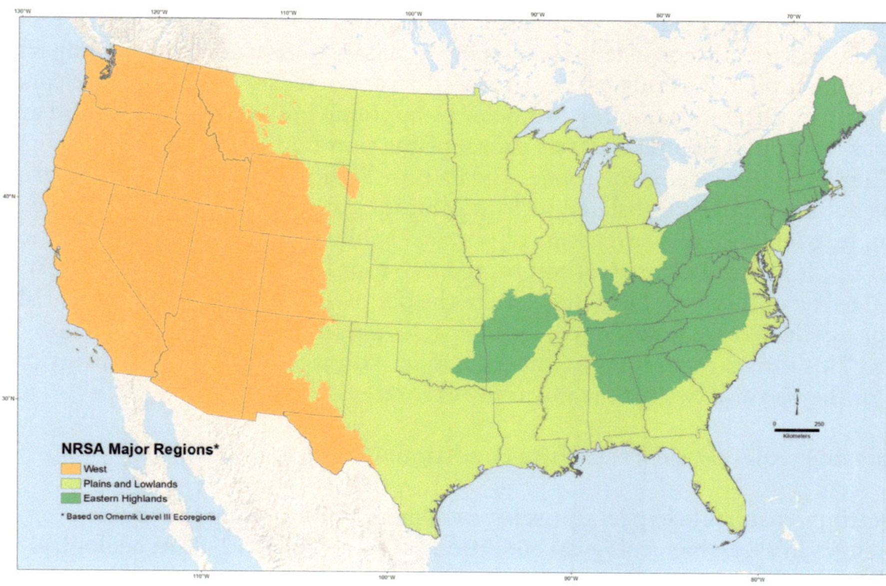

A

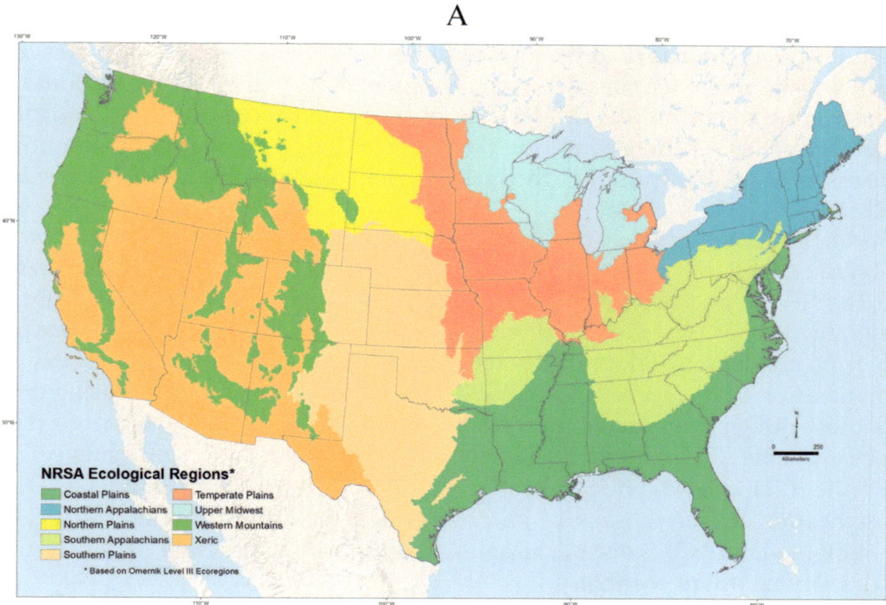

B

Figure 2.
(A) Three major landforms and climate reporting regions in the National Rivers and Streams Assessment (NRSA). (B) Nine aggregated ecoregions used for reporting in NRSA.

Fish and benthic macroinvertebrate samples, collected from each stream and river reach, are sent to taxonomists for identification [35, 36]. Water samples for chemical analyses are collected at mid-stream or river reach. Measurements of physical habitat attributes are collected at systematically spaced locations along the entire reach sampled. The chemical and physical habitat data are translated into descriptors of chemical or physical habitat or indicators of anthropogenic disturbance (i.e., stressors) that might impact biological condition.

The historic concerns about the lack of consistency and comparability in monitoring programs are resolved in the NRSA through the use of standardized field and laboratory protocols [32, 37]. Standardization allows the data to be combined to produce a nationally consistent assessment. Standardization also allows comparison to other methods. The 2004 survey provided an opportunity to examine the comparability of different sampling protocols by applying both the NRSA method and various state or USGS methods to a subset of the sites (e.g., [38, 39]).

The NRSA transforms the collected data into "indicators" that are meaningful to the public or can be translated into meaningful statements for the public. For example, over 3000 measurements of physical habitat structure are collected from each sample site and ultimately compacted into four indicators that can be meaningful to the public. Similarly, at each site the benthic macroinvertebrate and fish samples collected are reduced to a list of species present and their relative abundance. This information is then transformed into three indices of biotic integrity, one for the fish and two for the macroinvertebrates.

2.4 Setting expectations: reference conditions

Setting reasonable expectations for each indicator is among the greatest challenges in assessing ecological condition [40, 41]. For the NRSA, ecological condition assessments based on chemical, physical, and biological field measurements at each site were compared to a benchmark of what one would expect to find in relatively undisturbed streams and rivers within that region [42]. Sets of least disturbed reference sites within each region were used to: (1) develop and calibrate multimetric indices (MMIs) and observed/expected (O/E) indices, and (2) set thresholds for three condition classes: good, fair, and poor [42]. Conditions at these sets of relatively undisturbed stream and river sites are called "reference conditions".

Rather than relying solely on best professional judgment to set these reference condition benchmarks or even to finalize the sites considered least disturbed/reference, the NRSA data analysts first generated a pool of candidate sites that might potentially serve as least disturbed reference. Candidate sites for this reference pool came from either hand-selected sites recommended by State and EPA Regional participants or were screened as a subset from the pool of sites selected using the probability design site selection process. The only requirement was that site-specific data be available. This reliance on data for the final determination of reference sites rather than solely relying on best professional judgment as recommended in the application of Tiered Aquatic Life Use (TALU) framework and the biological condition gradient [43] is one of the hallmarks of NARS – the use of data-driven determinations where possible.

The pool of candidate reference sites was filtered through a set of physical and chemical data screens (i.e., riparian condition, nutrients, chloride, turbidity, excess fine sediments). When a site passed through all the data screens it was used to describe the distribution of condition indicators among least disturbed sites in that region (i.e., regional reference condition) "Pristine" landcover in watersheds was not required for a site to be considered "reference"; for example, sites in human-use dominated watersheds with local chemical and physical conditions among the best in the region could still be considered reference. The use of biological data for screening was avoided over concerns of circularity. For the same reason, physical habitat observations (e.g., riparian vegetation and streambed sediments) other than direct observations of human activities were not used to screen candidate reference sites for assessing physical habitat condition.

Not every reference site had identical chemical, physical, biological indicator scores. A range of values was found at the reference sites within an ecoregion. This range of values was used to construct a reference site distribution. The 5th and

25th (or 95th and 75th) percentiles of the reference-site distributions were used as thresholds for assigning any individual site in the probability survey to a condition class, i.e., good, fair, or poor.

2.5 Indicators of condition: biological quality

Samples of the macroinvertebrate and fish assemblages formed the basis for assessing the biological quality of streams and rivers. Only the macroinvertebrate assemblage results are presented here, although similar results are available for fish. Diatom assemblage samples were collected and analyzed and as of this writing, and taxonomic consistency issues are being resolved.

Two measures of the macroinvertebrate assemblage were used to communicate biological quality: a multimetric index (MMI) of macroinvertebrate integrity [10] and an observed/expected (O/E) index of taxa loss [11]. The MMI was developed for each of the nine aggregated ecoregions and compared with the reference conditions determined for that ecoregion [42].

O/E indices of taxa loss were also calculated. These are interpreted as the percentage of the expected taxa present at a site. Each tenth of a point less than 1 represents a 10% loss of taxa, e.g., an O/E value of 0.9 indicates 90% of the expected taxa are present and 10% are missing. Three O/E models were developed, one for each of the major climatic regions (**Figure 2A**): The Eastern Highlands, the Plains and Lowlands, and the West [11, 44]. Four categories of taxa loss were calculated: < 10% loss, 10–20% loss, 20–50% loss, and >50% taxa loss.

2.6 Indicators of stressors impacting streams and rivers

River and stream biota can be adversely impacted when alterations occur within the watershed or within the stream and river itself. The in-stream and riparian characteristics that are altered as a result of human activity and in turn result in biotic changes are considered "stressor indicators". These resulting aquatic stressors can be chemical [45], physical, or in some cases, biological [46]. Importantly, the goal of the CWA is to restore and maintain the chemical, physical, and biological integrity of the nation's water resources. The NRSA has a dual purpose in generating data on chemical, physical, and biological stressors. The first purpose uses these data in describing chemical and physical integrity of rivers and streams as a means of tracking progress toward the goals of the CWA. The second purpose uses these data to rank the stressors in their relative importance for policy. Ranking occurs in three ways. The first way establishes how widespread the stressors are. The second way ranks stressors by their severity when they occur, i.e., how likely are they to impact biota. And the third way, perhaps the most important, ranks stressors based on the likely improvement in rivers and streams if that stressor is reduced or eliminated. Not every potential chemical or physical stressor is currently included in the NRSA reports on condition, but both present and future surveys of rivers and streams in the US should include measurements that enable assessments of additional stressors for which there is reasonable concern that they may become important in the future.

The NRSA stressor indicators are the proximal stressors, i.e., changes in chemical or physical attributes that can affect biota. The stressors are not the more distal measures such as basin land-use or land-cover alterations not directly observed by the field crews, e.g., row crops, mining, or grazing visible in satellite imagery. This approach asserts that many human activities on the landscape can be sources of pollutants or indirect causes of stress to streams. However, the focus of the NRSA is to identify and quantify the stressors, rather than their sources. The general philosophy was to understand the most significant stressors first. This information can be used in the process of source tracking and determining probable causes, which are logical future steps for the NRSA and similar national assessments.

Eight stressor indicators were selected for reporting. Four stressors were chemical, and four were related to habitat alterations. The chemical stressors were excess total nitrogen (total N), excess total phosphorus (total P), excess salinity (based on conductivity), and acidification (based on acid neutralizing capacity). Prior 305(b) reports from States or national attention were the basis for these selections. Indicators of habitat alteration have not historically been included in monitoring by most water quality agencies. With a focus on the CWA goals, physical integrity became a needed element within NARS. Four indicators of physical integrity, excess fine sediments, alterations of instream fish habitat, alteration of riparian vegetation structure, and disturbance of the riparian zone are the initial focus. A fifth, hydrologic alteration is near completion.

2.7 Ranking of stressors: relative extent and relative risk

An important prerequisite to making policy and management decisions is an understanding of the relative magnitude or importance of potential stressors across a region and the expected benefit of reducing or eliminating that stressor. Both the prevalence (i.e., extent of stream length with high levels of the stressor) and the severity (i.e., impact on biological condition) of each stressor were considered. The NRSA reports include separate ranking for each of these elements, extent and risk.

Relative extent is a measure of how widespread the problem is...how much of the river and stream length has high levels of that particular stressor. Does high nitrogen occur in few or in many streams and rivers? Are high nitrogen levels geographically isolated or widespread? Relative risk, on the other hand, addresses the severity of the impact of high nitrogen on the biota when it occurs as compared to when nitrogen levels are low. Neither of these measures individually is a good indication that the problem should be addressed. But when combined, they provide powerful evidence of the need to act.

3. Results

Fish, macroinvertebrates and periphyton samples were all collected during the 2013–2014 stream and river survey. The data were processed and assessed and can be found in the detailed online dashboard and report [47]. Here we present the results for just the macroinvertebrate assemblage as an example of data generated by the NRSA.

3.1 Benthic macroinvertebrate conditions (MMI)

Nationally, 44% of the perennial stream and river length (hereafter simply referred to as "stream length") was in poor condition, and 26% was in fair condition as measured by the benthic macroinvertebrate MMI relative to the least-disturbed reference condition in each of the nine aggregated ecoregions (**Figure 3**). Based on the MMI, 42% of stream length in the Eastern Highlands, 47% of stream length in the Plains and Lowlands, and 31% of stream length in the West were in poor condition. Detailed examples of results for the nine aggregated ecoregions for the 2008–2009 NRSA are available elsewhere [28, 48].

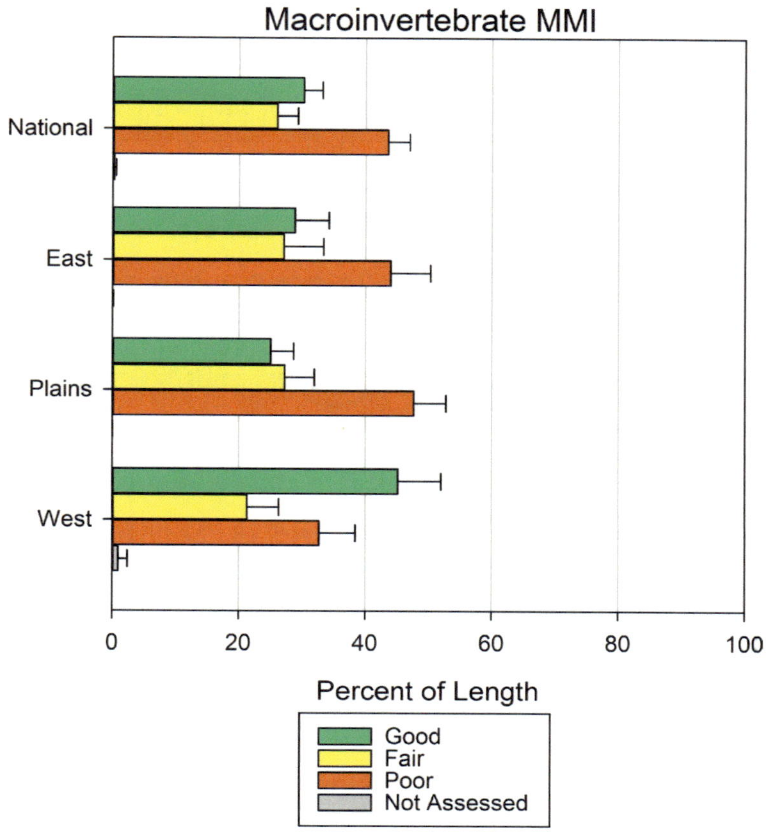

Figure 3.
National and regional results from the 2013–2014 National Rivers and Streams Assessment for the benthic macroinvertebrate multimetric index (MMI). Results are presented as the percent of stream length in good, fair and poor conditions, based on the degree of similarity to regionally-defined reference condition. Error bars represent 95% confidence intervals.

3.2 Benthic macroinvertebrate taxa (O/E index)

Nationally, 46% of stream length lost <10% of expected taxa, 13% lost 10–20%, 26% of stream length lost 20–50%, and 15% of stream length lost >50% of expected taxa (**Figure 4**). The Eastern Highlands experienced the greatest loss of expected taxa; 21% of stream length lost >50%, 29% of length lost 20–50% of expected taxa, 10% of length lost 10–20% of taxa, and 40% of stream length lost <10% of expected taxa.

3.3 Relative extent of stressors

High levels of several stressors occurred throughout perennial streams and rivers. Excess total phosphorus was the most widespread stressor nationally and within each region. Fifty-eight percent of the river and stream length are marked by high total phosphorus concentrations across the country (**Figure 5A**). The prevalence in the Plains and Lowlands, Eastern Highlands and the West is 51, 73 and 49%, respectively.

Nutrients (total phosphorus and total nitrogen) were consistently the most extensively occurring stressors with the stream length in poor condition ranging

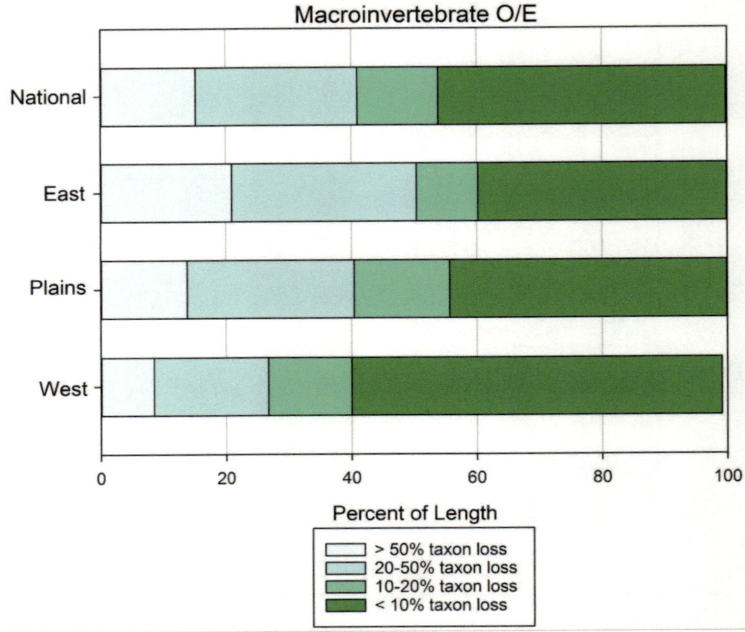

Figure 4.
National and regional results from the 2013–2014 National Rivers and Streams Assessment for the benthic macroinvertebrate observed/expected (O/E) index of taxon loss. Results are presented as the percent of stream length in four categories of taxon loss.

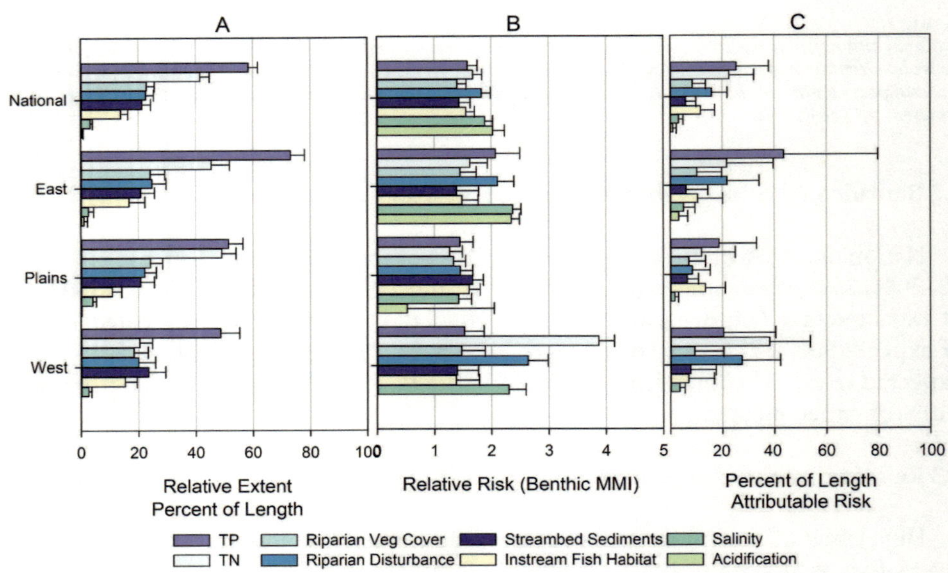

Figure 5.
Relative ranking of stressors nationally and regionally for the 2013–2014 National Rivers and streams assessment. (A) Relative extent is the percent of stream length in poor condition for each of the eight stressors evaluated. (B) Relative risk of observing poor biological condition (based on values of the benthic invertebrate multimetric index [MMI]) given poor stressor conditions relative to observing poor MMI values given good or moderate stressor conditions. (C) Attributable risk is the percent of improvement (i.e., decrease) in stream length in poor biological condition (based on MMI scores) given that a stressor level is modified from poor to good or fair condition. Error bars represent 95% confidence intervals.

from about 20% to about 75% across the three major regions (**Figure 5A**). Poor conditions for the four physical habitat indicators were observed in about 20% of stream length nationally, but ranged from 10 to 25% across the three climatic regions. There was much more variability in physical habitat condition at the finer ecoregion scale, with 4 to 40% of stream length in poor condition among the 9 ecoregions, depending on the specific physical habitat indicator and region. Alteration of riparian vegetation cover was the most extensive habitat stressor nationally and in the Eastern Highlands and the Plains and Lowlands regions. High levels of excess fine sediments were most prevalent in the West.

3.4 Relative risk of stressors

Almost all stressors evaluated in the NRSA were associated with increased risk for poor macroinvertebrate condition (**Figure 5B**). Nationally, the relative risk values ranged from 1.4–2.0, with only slight or no substantial difference among the stressors nationally. In fact, two of the stressors, acidification and increased salinity, had among the largest relative risk values.

Relative risk values differed among major NRSA regions (**Figure 5B**). The largest relative risk value (3.9) occurred for total nitrogen in the West, showing that streams with excess total nitrogen were nearly 4 times more likely to have their benthic macroinvertebrate assemblage in poor condition when compared to streams with moderate or low concentrations of total nitrogen. All the stressors posed a risk to macroinvertebrate biological integrity with relative risks values ranging from 1.3 to 3.9 nationally and in all three geoclimatic regions.

3.5 Attributable risk - combining stressor extent and relative risk

As described above, the use of relative extent and relative risk in combination provides the best assessment of a particular stressor. It provides an estimate of the relative improvement in the biota with the reduction of that stressor (**Figure 5A–C**). Rivers and streams are at greatest risk when the stressor is both widespread (large percentage of river and stream length with stressor at excess levels, **Figure 5A**) and presents potentially severe effects (i.e., high relative risk values, **Figure 5B**). Another tool from epidemiology, the concept of attributable risk, was adapted and applied to the data from the Wadeable Streams Assessment [49], and is now part of all of assessments produced from the NRSA surveys. Attributable risk combines relative extent with relative risk to produce a single number that can be used to rank stressors and to inform management decisions by suggesting the level of improvement expected (in terms of the % of stream length in poor biological condition that could be elevated to good condition) if excess levels of a particular stressor are reduced to moderate or low levels.

Nationally, excess total nitrogen and total phosphorus are the stressors whose relative extent (how widespread) and relative risk (severity of impact when excess levels occurred) suggest the largest expected improvement. For each of these nutrients, roughly a 25% improvement (i.e., decrease) in the stream length in poor biological condition is expected if levels of these nutrients are reduced from excess to moderate or low (**Figure 5C**). Excess fine sediments and alteration of the riparian vegetation were the habitat stressors that would produce the largest expected improvement in stream and river biological condition (a 16 and 12% improvement, respectively). Salinity occurs in excess levels in a very low percentage of stream length (**Figure 5A**) and despite high relative risk (**Figure 5B**), this stressor has a very small attributable risk. Thus, excess salinity might be considered a local issue requiring a local targeted management approach, severe when it occurs, yet not of significance at a national or regional scale.

4. Conclusions

The NRSA surveys began in early 2000s and were repeated in 2008–2009, 2013–2014 and most recently in 2018–2019. The results of the NRSA and the data on which they are based constitute a baseline from which future trends can be evaluated. The NRSA survey has been repeated enough that detecting changes and trends in status are now possible using the NRSA approach. Stoddard et al. [50] demonstrated the NRSA's capability for detecting changes and trends when they reported a consistent increase in total phosphorus concentration and a loss of low nutrient waters across surveys in the period of 2004 and 2014. As the number of resurveys mounts up over time, results from trend detection and analyses will increase, becoming a more and more critical contribution of the NRSA results and the NARS in general.

Although the set of important stressors currently assessed by NRSA appears robust for long-term trends in important known stresses on biological integrity, there is room for innovation and inclusion of new and relevant indicators of stress. There is also room for integration of new monitoring technologies such as DNA sequencing, LIDAR and new satellite-based sensor technology.

The NRSA was the first and is still the only comprehensive national assessment of water resources conducted in the US that is based on uniform, consistent field protocols and a statistically robust sampling design. The NRSA statistical design is a major advancement in aquatic monitoring and has been embraced by multiple States and Federal Agencies. The NRSA statistical design and many NRSA field sampling methods and analytical approaches have been applied or adapted to monitoring and assessment within US states and worldwide (Canada, Brazil, Bolivia, Belize, and China). The CWA goals of restoring and maintaining the chemical, physical and biological integrity of the Nation's waters imply that we would have the required monitoring to track our progress toward meeting those goals. The NRSA and the other surveys within the NARS, as well as those States and other agencies adopting the NARS tools, are beginning to deliver on that implicit promise.

Acknowledgements

We gratefully acknowledge the tremendous effort and collaboration on the part of the USEPA Office of Water, the USEPA Office of Research and Development, and numerous State, Tribal, federal, and contractor organizations that provide field, laboratory, and data analysis support for the National Rivers and Streams Assessment. This research was performed while ATH held a National Research Council Senior Research Associateship award at the USEPA Office of Research and Development, Pacific Ecological Systems Division, Corvallis, Oregon.

We appreciate the constructive reviews of earlier drafts by L. Herger, R. Cook, and J. Markwiese. This chapter has been subjected to Agency review and has been approved for publication. The views expressed in this paper are those of the author(s) and do not necessarily reflect the views or policies of the U.S. Environmental Protection Agency. Any mention of trade names, products, or services does not imply an endorsement by the U.S. Government or the U.S. Environmental Protection Agency. The EPA does not endorse any commercial products, services, or enterprises.

This work was authored by United States Government employees as part of their official duties. In view of Section 105 of the Copyright Act (17 U.S.C. §105) the work is not subject to U.S. copyright protection.

Conflict of interest

The authors declare no conflict of interest.

Author details

Steven G. Paulsen[1]*, David V. Peck[1], Philip R. Kaufmann[1,2] and Alan T. Herlihy[2,3]

1 Office of Research and Development, US Environmental Protection Agency, Corvallis, Oregon, USA

2 Department of Fisheries and Wildlife, Oregon State University, Corvallis, Oregon, USA

3 National Research Council Senior Fellow, Pacific Ecological Systems Division, Office of Research and Development, US Environmental Protection Agency, Corvallis, OR, USA

*Address all correspondence to: paulsen.steve@epa.gov

References

[1] Shapiro MH, Holdsworth SM, Paulsen SG. The need to assess the condition of aquatic resources in the US. Journal of the North American Benthological Society. 2008;**27**(4):808-811

[2] Harris G. Inland Waters Theme Commentary Prepared for the 2006 Australian State of the Environment Committee, Department of the Environment and Heritage, Canberra, Australia. 2006. Available from: http://www.deh.gov.au/soe/2006/commentaries/water/index.html

[3] State of the Environment Advisory Council. Australia: State of the Environment 1996. An Independent Report Presented to the Commonwealth Minister for the Environment. Collingwood, Victoria Australia: CSIRO Publishing; 1996. 502 + 47 p

[4] Ball J, Donnelly P, Erlanger P, Evans R, Kollmorgon A, Neal B, et al. Inland Waters. Australia State of the Environment Report 2001 (Theme Report). CSIRO, Department of the Environment and Heritage: Canberra, Australia; 2001

[5] Hering D, Borja A, Carstensen J, Carvalho L, Elliott M, Feld CK, et al. The European Water Framework Directive at the age of 10: A critical review of the achievements with recommendations for the future. Science of the Total Environment. 2010;**408**(19):4007-4019

[6] Hering D, Verdonschot PFM, Moog O, Sandin L. Preface. Hydrobiologia. 2004;**516**(1):7-9

[7] Norris RH, Morris KR. The need for biological assessment of water quality: Australian perspective. Australian Journal of Ecology. 1995;**20**(1):1-6

[8] Simpson JC, Norris RH. Biological assessment of river quality: Development of AUSRIVAS models and outputs. In: Wright JF, Sutcliffe DW, Furse MT, editors. Assessing the Biological Quality of Freshwaters: RIVPACS and Other Techniques. Ambleside, Cumbria, U.K.: Freshwater Biological Association; 2000. pp. 125-142

[9] Hering D, Moog O, Sandin L, Verdonschot PFM. Overview and application of the AQEM assessment system. Hydrobiologia. 2004;**516**:1-20

[10] Stoddard JL, Herlihy AT, Peck DV, Hughes RM, Whittier TR, Tarquinio E. A process for creating multimetric indices for large-scale aquatic surveys. Journal of the North American Benthological Society. 2008;**27**(4):878-891

[11] Yuan LL, Hawkins CP, Van Sickle J. Effects of regionalization decisions on an O/E index for the US national assessment. Journal of the North American Benthological Society. 2008;**27**(4):892-905

[12] Olsen AR, Peck DV. Survey design and extent estimates for the Wadeable streams assessment. Journal of the North American Benthological Society. 2008;**27**(4):822-836

[13] Paulsen SG, Hughes RM, Larsen DP. Critical elements in describing and understanding our nation's aquatic resources. Journal of the American Water Resources Association. 1998;**34**(5):995-1005

[14] Federal Water Pollution Control Act - Amendments of 1972., Pub. L. No. 33 U.S.C. §1251 et seq. (1972)

[15] National Research Council. Environmental Monitoring, Volume IV. Washington, DC: National Academy of Sciences; 1977

[16] National Research Council. Linking Science and Technology to Society's Environmental Goals. Washington, DC: The National Academies Press; 1996. p. 544

[17] GAO (Government Accountability Office). Better monitoring techniques are needed to assess the quality of rivers and streams. Volume 1. Washington, DC: Community, and Economic Development Division, Government Accountability Office; 1981. Report No.: GAO/CED–81-30

[18] GAO (Government Accountability Office). Water quality: key EPA and State decisions limited by inconsistent and incomplete data. Gaithersburg, Maryland: Resources, Community, and Economic Development Division, Government Accountability Office; 2000. Report No.: GAO/RCED–00-54

[19] GAO (U.S. General Accounting Office). Protecting human health and the environment through improved management. Gaithersburg, Maryland: Resources, Community, and Economic Development Division, General Accounting Office; 1988. Report No.: GAO/RCED–88-101

[20] USEPA (United States Environmental Protection Agency). Mercury in solids and solutions by thermal decomposition, amalgamation, and atomic absorption spectrometry. Draft Method 7473. Washington, DC: EPA; 1998

[21] H. John Heinz Center for Science, Economics, the Environment. The State of the Nation's Ecosystems: Measuring the Lands, Waters, and Living Resources of the United States. New York: Cambridge University Press; 2002. p. 288

[22] Environmental Mysteries. USA Today. 2002 September 26. 2002

[23] Stoddard JL, Peck DV, Paulsen SG, Van Sickle J, Hawkins CP, Herlihy AT, et al. An Ecological Assessment of Western Streams and Rivers. Washington, DC, Office of Research and Development, US Environmental Protection Agency; 2005 Report No.: EPA 620/R-05/005

[24] Paulsen SG, Mayio A, Peck DV, Stoddard JL, Tarquinio E, Holdsworth SM, et al. Condition of stream ecosystems in the US: An overview of the first national assessment. Journal of the North American Benthological Society. 2008;**27**(4):812-821

[25] US Coast Guard. Federal Requirements for Recreational Boats. Washington, DC: U.S. Department of Transportation, United States Coast Guard; 1987

[26] USEPA (United States Environmental Protection Agency). National Rivers and Streams Assessment 2008-2009 Technical Report. Washington, DC: Office of Water and Office of Research and Development, US Environmental Protection Agency; 2016. Report No.: EPA 841/R-16/008

[27] McKay L, Bondelid T, Dewald TG, Johnston J, Moore R, Rea A. NHDPlus Version 2: User Guide (Data Model Version 2.2, updated March 13, 2019). Washington, DC: U.S. Environmental Protection Agency, Office of Water; 2012

[28] USEPA (United States Environmental Protection Agency). National Rivers and Streams Assessment 2008-2009: A Collaborative Survey. Washington, DC: Office of Water and Office of Research and Development, US Environmental Protection Agency; 2016. Report No.: EPA 841/R-16/007

[29] USEPA (United States Environmental Protection Agency). Visiting a Regional Office webpage. Washington, DC: U.S. Environmental Protection Agency; 2020; Available

from: https://www.epa.gov/aboutepa/visiting-regional-office [Accessed: 13 March 2020]

[30] USEPA (United States Environmental Protection Agency). National Rivers and Streams Assessment. A Collaborative Survey. Washington, DC: Office of Water and Office of Research and Development, US Environmental Protection Agency; 2013-2014 in press

[31] Hughes RM, Peck DV. Acquiring data for large aquatic resource surveys: The art of compromise among science, logistics, and reality. Journal of the North American Benthological Society. 2008;**27**(4):837-859

[32] USEPA (United States Environmental Protection Agency). National Rivers and Streams Assessment 2013/14: Field Operations Manual - Wadeable. Washington, DC: Office of Water and Office of Environmental Information; 2013. Report No.: EPA 841/B-12/009b

[33] USEPA (United States Environmental Protection Agency). National Rivers and Streams Assessment 2013/14: Field Operations Manual - Non-Wadeable. Washington, DC: Office of Water and Office of Environmental Information; 2013. Report No.: EPA 841/B-12/009a

[34] USEPA (United States Environmental Protection Agency). National Rivers and Streams Assessment 2013-14: Quality Assurance Project Plan, v. 2.1. Washington, DC: U.S. Environmental Protection Agency, Office of Water; 2015. Report No.: EPA 841-B-12-007

[35] Stribling JB, Pavlik KL, Holdsworth SM, Leppo EW. Data quality, performance, and uncertainty in taxonomic identification for biological assessments. Journal of the North American Benthological Society. 2008;**27**(4):906-919

[36] USEPA (United States Environmental Protection Agency). National Rivers and Streams Assessment 2013-2014: Laboratory Operations Manual (Version 2.0). Washington, DC: Office of Water; 2014. Report No.: EPA-841-B-12-010

[37] USEPA (United States Environmental Protection Agency). Wadeable Streams Assessment: Field Operations Manual. Washington, DC: Office of Water; 2004. Report No.: EPA 841/B-04/004

[38] Carlisle DM, Hawkins CP. Land use and the structure of western US stream invertebrate assemblages: Predictive models and ecological traits. Journal of the North American Benthological Society. 2008;**27**(4):986-999

[39] Ode PR, Hawkins CP, Mazor RD. Comparability of biological assessments derived from predictive models and multimetric indices of increasing geographic scope. Journal of the North American Benthological Society. 2008;**27**(4):967-985

[40] Stoddard JL, Larsen DP, Hawkins CP, Johnson RK, Norris RH. Setting expectations for the ecological condition of streams: The concept of reference condition. Ecological Applications. 2006;**16**:1267-1276

[41] Hawkins CP, Olson JR, Hill RA. The reference condition: Predicting benchmarks for ecological and water-quality assessments. Journal of the North American Benthological Society. 2010;**29**(1):312-343

[42] Herlihy AT, Paulsen SG, Sickle JV, Stoddard JL, Hawkins CP, Yuan LL. Striving for consistency in a national assessment: The challenges

of applying a reference-condition approach at a continental scale. Journal of the North American Benthological Society. 2008;**27**(4):860-877

[43] Davies SP, Jackson SK. The biological condition gradient: A descriptive model for interpreting change in aquatic ecosystems. Ecological Applications. 2006;**16**(4):1251-1266

[44] USEPA (United States Environmental Protection Agency). Wadeable Streams Assessment: A Collaborative Survey of the Nation's Streams. Washington, DC: Office of Water, US Environmental Protection Agency; 2006. Report No.: EPA 641/B-06/002

[45] Herlihy AT, Sifneos J. Developing nutrient criteria and classification schemes for wadeable streams in the conterminous USA. Journal of North American Benthological Society. 2008;**27**(4):932-948

[46] Ringold PL, Magee TK, Peck DV. Twelve invasive plant taxa in US western riparian ecosystems. Journal of the North American Benthological Society. 2008;**27**(4):949-966

[47] USEPA. National Rivers and Streams Assessment Webpage. Washington, DC: U.S. Environmental Protection Agency; 2018; Available from: https://www.epa.gov/national-aquatic-resource-surveys/nrsa

[48] USEPA. National Rivers and Streams Assessment 2008-2009 Results Webpage. Washington, DC: U.S. Environmental Protection Agency. p. 2018; Available from: https://www.epa.gov/national-aquatic-resource-surveys/national-rivers-and-streams-assessment-2008-2009-results

[49] Van Sickle J, Paulsen SG. Assessing the attributable risks, relative risks, and regional extents of aquatic stressors. Journal of the North American Benthological Society. 2008;**27**(4):920-931

[50] Stoddard JL, Van Sickle J, Herlihy AT, Brahney J, Paulsen S, Peck DV, et al. Continental-scale increase in lake and stream phosphorus: Are oligotrophic systems disappearing in the United States? Environmental Science & Technology. 2016;**50**(7):3409-3415

Chapter 8

Lessons Learned from 30 Years of Assessing U.S. Coastal Water

John A. Kiddon, Hugh Sullivan, Walter G. Nelson, Marguerite C. Pelletier, Linda Harwell, Mari Nord and Steve Paulsen

Abstract

The 1972 Clean Water Act (CWA) established goals and regulations regarding water quality in the U.S. water resources, including coastal waters. The U.S. Environmental Protection Agency (EPA) was charged with implementing the CWA's goals and with helping states, and tribes meet their mandate to periodically monitor and assess water quality in their jurisdictions. In response, the EPA initiated the Environmental Monitoring and Assessment Program (EMAP) to develop and test effective methods of assessing water quality in lakes, rivers and streams, and estuaries at state and national scales. EMAP-Estuaries commenced in 1990, devising sampling designs and protocols for estuaries, testing potential indicators, establishing assessment, and reporting methods. Estuarine research and development efforts continued in a series of subsequent programs, each adapting and adopting the best practices of earlier programs, each becoming more national in scale, and each integrating state and tribal participation to a greater degree. Recent surveys have included an assessment of coastal Great Lakes waters. This chapter recounts the history of assessments in coastal waters, emphasizing the current approach while highlighting examples of lessons learned over the 30-year development period leading to the National Coastal Condition Assessment.

Keywords: coastal assessment, EMAP, NCA, NCCA, NARS, indicators

1. Introduction

The 1960s were a decade of growing awareness and concern regarding the declining quality of the surface waters of the U.S., most dramatically exemplified when the Cuyahoga River, Ohio caught fire in the summer of 1969. In response, the U.S. Congress passed the Clean Water Act (CWA) in 1972 [1], establishing goals and regulations governing the restoration and maintenance of the nation's water resources, including coastal regions. The CWA also specifically addressed the need for monitoring water quality. Section 305b of the CWA required states and tribes to survey and periodically report on the overall condition of their surface waters, including coastal waters. In addition to the state programs, numerous other water quality monitoring and research programs were initiated in major estuarine systems, such as Chesapeake Bay, Narragansett Bay, Tampa Bay, and Puget Sound.

However, for the first two decades of the Act, reviewers consistently highlighted the fact that the approaches used by the states and tribes to monitor conditions were not nationally consistent and the information they reported could not be consolidated into a single assessment of the Nation's waters [2–7]. Despite substantial expenditures, regulators were unable to judge the effectiveness of pollution-control legislation [8]. In response to these limitations, the EPA initiated the Environmental Monitoring and Assessment Program (EMAP), a research effort that spanned 17 years. These EMAP efforts would eventually evolve into what is now known as EPA's National Aquatic Resource Surveys (NARS) which continues to optimize approaches to conducting large-scale water quality assessments in lake, river, stream, estuarine and wetland resources across the U.S. This chapter focuses on the estuarine components of the EMAP and NARS assessments. An overview of EPA's efforts to assess coastal waters is presented in **Figure 1**. The timeline can be divided into three phases.

Beginning in 1990 and continuing for a decade, a series of regional assessments were executed in the major U.S. coastal ecological provinces. These EMAP-Estuaries programs explored innovative methods of conducting coastal assessments and established several of the defining features of EPA's assessment approach. For instance, EMAP planners adopted probabilistically-derived survey designs that minimized sampling bias, and designated sites that were appropriately weighted to estimate—with confidence intervals—the percentage of a region in good, fair, or poor condition. The early programs also developed a common core of indicators that could be used regionally or nationally to characterize conditions in key components of estuarine ecosystems—the water column, sediment, and benthic and fish communities. The lessons of these efforts were reported in many technical statistical summaries and summary reports, e.g., [9–14], but relatively few of the accounts were prepared with the public reader in mind. This research and developmental phase was led by EPA's Office of Research and Development (ORD) in partnership

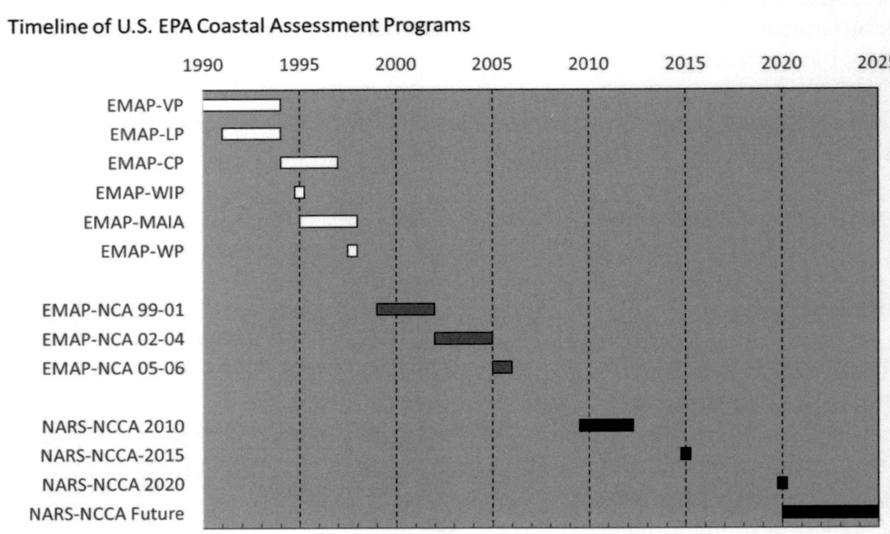

Figure 1.
EPA coastal assessment programs—Development and implementation phases. EMAP, Environmental Mapping and Assessment Program; Regional development: VP, Virginian Province (U.S. NE Atlantic coast); LP, Louisianian Province (Gulf of Mexico coast); CP, Carolinian Province (U.S. SE Atlantic coast); WIP, West Indian Province (South Florida coast); MAIA, Mid-Atlantic Integrated Assessment (Chesapeake, Delaware & Albemarle-Pamlico bays); WP, Western pilot (U.S. Pacific coast); NCA, National Coastal Assessment; Nationwide development phase NCCA, National Coastal Condition Assessment; Nationwide implementation phase.

with other federal agencies, especially the National Oceanic and Atmospheric Agency (NOAA) and U.S. Fish and Wildlife Service (FWS), and with some participation of state environmental agencies and academic institutions.

In the second phase, the EMAP-Estuaries program expanded nationally into the National Coastal Assessment, NCA 1999–2006. The NCA was also an EMAP research program, with primary goals of adopting and refining the best techniques developed in the regional studies and applying them to conduct coastal assessments at both national and regional scales [15]. The NCA evaluations continued the approach of assessing four key aspects of estuarine ecosystems, i.e., water quality, sediment quality, and the ecological condition of benthic and fish communities. Of equal importance, the NCA worked to more fully engage the states and tribes in the assessment process; thereby facilitating compliance with Section 305b of the Clean Water Act. As information accrued, the NCA also experimented with ways of analyzing and reporting how coastal conditions changed over time. Four National Coastal Condition Reports (NCCR I–IV) resulted from these efforts [16–19]. Particular attention was paid to explaining the assessment process and results to the general public.

After 16 years of research, development, and stakeholder feedback, the coastal monitoring approach was deemed ready for routine deployment, and responsibility for implementation was passed from EPA's ORD to EPA's Office of Water (OW). Now renamed as the National Coastal Condition Assessment (NCCA), surveys were executed in 2010 and 2015, and plans are underway to conduct assessments in 2020 and beyond. Beginning with the 2010 survey, the coastal waters of the Great Lakes were included as part of the NCCA program despite the substantial differences in the freshwater and estuarine realms [20]. The NCCA, together with the National Lakes Assessment (NLA), the National Rivers and Streams Assessment (NRSA), and the National Wetlands Condition Assessment (NWCA), form the EPA National Aquatic Resources Surveys (NARS) program [21]. The goals of NARS are (i) to conduct routine surveys of all surface-water resources of the U.S. on a regular schedule; (ii) issue reports on assessments of each resource; and (iii) establish a joint database useful for conducting assessments and modeling investigations concerning all components of the surface-water systems.

In short, the EPA and its partners have devised an ambitious and unique approach of conducting multi-scale ecological assessments of the nation's coastal waters. NCCA and NARS reflect the results of concerted research and a pragmatic willingness to modify techniques and protocols based on lessons learned. Although logistically challenging, incorporating states and tribes in all aspects of the surveys has proved to be a clear success, both by enhancing the assessments and, more importantly, by helping build capacity of the states and tribes to conduct surveys on their own. Finally, the programs provide useful metrics by which environmental managers and legislators could judge the effectiveness of implemented policies. The remainder of this chapter further describes EPA's approach to assessing coastal waters, focusing primarily on the methods employed in the recent NCCA 2010 and 2015 surveys, which are the most thoroughly documented programs. Significant differences from earlier or later surveys are highlighted to emphasize how evolution shaped EPA's assessment process. Furthermore, where assessment approaches are similar in estuaries and the Great Lakes, we focus on the estuarine methodology in deference to brevity. Full documentation, data, and reports concerning both estuarine and Great Lakes assessments are available at [22]. The intended audiences for this chapter are knowledgeable scientists and environmental managers interested in reviewing the unique coastal assessment methods developed over 30 years of experimentation.

2. Key features of EPA's coastal assessments

During the summer of 2010, nearly 50 field crews visited 1104 pre-selected sampling stations in U.S. estuaries and Great Lakes coastal waters. Onsite, the crews collected environmental data and sampled the water column, sediments, and benthic and fish communities. Preserved samples were shipped to a dozen or so laboratories for analysis, and laboratory and field data were ultimately compiled into databases for analysis and reporting. In the following sub-sections, we outline the NCCA procedures used to select sampling stations, collect samples and information onsite, and assess and report ecological conditions at various scales. Further details regarding the implementation and evolution of assessment methods are described in Section 3.

2.1 Selecting sites

The NCCA employed a rigorous design process to meet several key assessment goals. First, the coastal waters to be assessed—the *target population*—was precisely specified. The target population was carefully defined as: (i) all estuarine waters in the conterminous U.S. from the "head-of-salt" (landward extent of waters with salinity greater than 0.5 ppm) to the boundary with the open ocean and (ii) Great Lakes of the U.S nearshore coastal waters located within 5 km of shore and less than 30 m in depth. EMAP included some inland river sections, river mouths, tidal streams and ponds, and sections of the continental shelf. However, NCCA excluded such waters as tidal streams and deep central channels of major rivers and bays, non-estuarine shorelines to better accommodate state needs. A GIS file specifying the areal coverage of the study regions—the *sample frame*—was compiled for subsequent use in selecting sampling sites. The sample frame for the 2010 NCCA comprised an area of 91,700 sq. km of coastal marine and fresh water.

Next, sampling sites were selected using a probabilistic, stratified survey design. That is, the target region was divided into *strata*—multiple nested assessment units that fairly represented the nation and various subregions designated for evaluation [23]. The largest domains were the five reporting regions: (1) The Atlantic coast of the northeastern U.S. from Maine through Virginia (including Chesapeake Bay); (2) the Atlantic coast of the southeastern U.S. from North Carolina through Biscayne Bay in south Florida; (3) the Gulf of Mexico coast of the U.S. from Biscayne Bay through Texas; (4) the Pacific coast of the western U.S. from Washington through California; and (5) the U.S. coasts of the Great Lakes. These strata were subdivided to delineate the coastal waters of the 21 ocean states and the five Great Lakes, and some larger units were in turn further subdivided to highlight important water bodies or regions designated for special study. The 2010 NCCA sampling design recognized a total of 64 distinct strata, which could be combined as needed for analysis or reporting at multiple spatial scales. Survey planners specified the number of sites allotted to each stratum based on cost effectiveness and survey priorities. The five primary reporting regions and the distribution of sampling sites in the 2010 NCCA survey are shown in **Figure 2**.

Sample sites were then selected probabilistically, but not randomly, using a process termed the Generalized Random Tessellation Survey (GRTS) design [23]. The GRTS method employs an intricate algorithm that ensures uniform and unbiased station placement, thereby minimizing clumping that may result if sites were selected using a purely randomized approach. A weighting factor, called the *inclusion probability*, was provided for each site. The factor was calculated as the stratum area divided by the number of sites in the stratum and was used during the analysis stage to estimate regional condition (see Section 2.3).

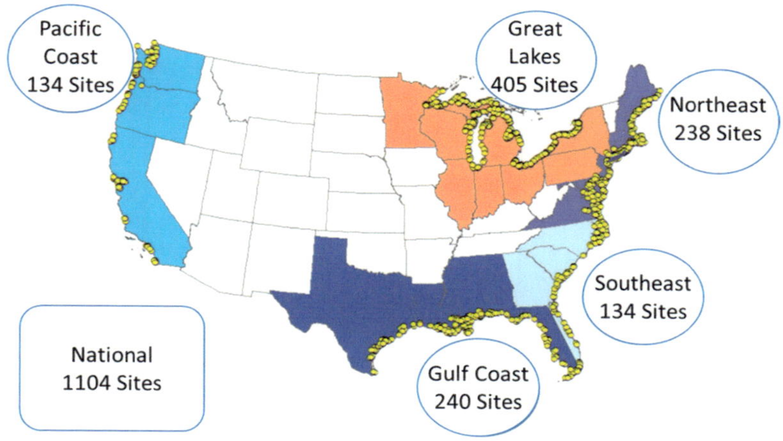

Figure 2.
Location of the 1104 sites sampled in the 2010 NCCA survey, by reporting region.

Finally, the survey design procedure identified both "base" and "oversample" locations. Sampling was mandatory at the base sites, and oversample sites were designated as replacements for inaccessible base sites or to be used by states in other regional assessments or enhancements. At least 50 base sites were allocated to each of the five reporting regions and to strata receiving an enhanced assessment. A sample size of 50 sites was considered adequate to yield results with reasonable statistical confidence [23]. Ten percent of the base sites were designated as "revisit sites", to be sampled twice during the same summer period in order to estimate intra-site variability. Additionally, 25% of sites were identified as "return sites"—stations to be repeatedly reassessed over the course of four subsequent NCCA surveys. These return sites increase the ability to quantify temporal variance and to aid in detecting change over time [24]. Further details of the entire NCCA site selection process is available in a non-technical overview of monitoring design topics provided online [25].

2.2 Implementing coastal surveys

The 2010 NCCA survey was a highly orchestrated campaign mounted to assess the nation's coastal waters. Implementation included training field crews, documenting sampling and analysis methods, collecting information and physical samples onsite, coordinating sample analysis, building databases, and performing quality assurance (QA) reviews.

Nearly 50 field crews composed of state, tribal, EPA personnel, and contractor staff, were deployed to collect samples and information during a summer *index sampling period*—June through September. Prior to the field season, the crews were rigorously trained by EPA trainers regarding NCCA protocols stipulated in the Site Evaluation Manual, Field Operations Manual, and Quality Assurance Project Plan [22]. During time on station, field crews would (i) record field conditions, including Secchi depth, vertical profiles of temperature, salinity, pH, dissolved oxygen, and photosynthetically active radiation (PAR) intensity; (ii) collect surface water samples for lab analysis of nutrients, chlorophyll, and human health indicators; (iii) collect grab sediment samples for analysis of contaminant concentrations, grain size, toxicity, and total organic carbon; (iv) collect and preserve separate sediment samples for characterization of the benthic macroinvertebrate community; and (v) collect fin-fish from within a proscribed distance from the site to characterize the local fish community and provide tissue for analysis of lipid and contaminant content (**Table 1**).

National Coastal Condition Assessment Indicators	
Physical habitat parameters	**Ecological contaminants in sediments and fish tissue**[a]
Physical habitat parameters Temperature (°C) Salinity (ppt) Dissolved oxygen—DO (mg/L) pH Secchi depth (m) Total organic carbon—TOC (%) % Silt/clay (grainsize) **Water-quality parameters** Chlorophyll *a* (µg/L) Ammonium (mg N/L) Nitrate plus nitrite (mg N/L) Nitrite (mg N/L) Dissolved inorganic nitrogen—DIN (mg N/L) Total nitrogen—TN (mg N/L) Dissolved inorganic phosphate—DIP (mg P/L) Total phosphorus—TP (mg P/L) **Sediment toxicity** Amphipod survival bioassay Estuarine test organisms: *Leptocheirus plumulosus* *Eohaustorius estuarius* Great Lakes test organism: *Hyalella azteca*	**Metals (µg/g)**: Ag, Al, As, Cd, Cr, Cu, Fe, Hg, Mn, Ni, Pb, Se, Sn, Tl, Zn **PAHs (ng/g)**: acenaphthene, acenaphthylene, anthracene, benz(a) anthracene, benz(a,e)pyrene, benzo(b,k,b+k)flouranthene, benzo(g,h,i)perylene, biphenyl, chrysene, dibenz(a,h) anthracene, fluoranthene, fluorene, indeo(1,2,3-c,d)pyrene, naphthalene, 1-methylnaphthalene, 2-methyl naphthalene, 2,6-dimethylnaphthalene, 2,3,5-trimethylnaphthalene, perylene, phenanthrene, 1-methylphenanthrene, pyrene, total PAHs **PCB congeners (ng/g)**: 8, 18, 28, 29, 44, 52, 66, 87, 101, 105, 118, 128, 138, 153, 170, 180, 187, 195, 201, 206, 209, total PCBs **Pesticides (ng/g)**: aldrin, chlordane (alpha-, gamma-, oxy-), dieldrin, dibenzothiophene, DDD (2,4'; 4,4'), DDE (2,4'; 4,4'), DDT (2,4'; 4,4'), endosulfan I & II, endrin, heptachlor, heptachlor epoxide, hexachlorobenzene, hexachlorohexane (alpha-, beta-, delta-), lindane, mirex, trans-nonachlor
Biotic conditions Diversity and abundance of benthic macroinvertebrates Diversity and abundance of fish	
Human health indicators	
Mercury in fish plugs; algal toxins (microcystin and cylindrospermopsin) and enterococcus in water	
PCBs, PBDEs and PFCs in fish tissue (Great Lakes only)	

[a]*National Oceanic and Atmospheric Administration National Status and Trends Program analytes. Concentrations reported as dry weight of sediments and wet weight of tissue.*

Table 1.
Indicators measured in the National Coastal Condition Assessment (NCCA) surveys.

The field data were submitted as either physical or electronic data sheets to NCCA headquarters for compilation. Preserved water, sediment, and fish samples were shipped to approve national or state laboratories for analysis and results were submitted to NCCA headquarters. Each site generated hundreds of field and laboratory data values that were organized into files by type (e.g., field data, water quality data, benthic census data, etc.), and maintained as "raw files" in a centralized database by information management specialists. The raw files were then subjected to a stringent two-phase QA review process, first checking for basic compliance with submission requirements (e.g., proper units, range checks, and conformity with standard taxonomic terminology). Any revisions to the raw files were carefully documented, and finalized files were made available at the NCCA public website [26].

One of the hallmarks of the NCCA, and the NCA which preceded it, has been the emphasis on the cooperation and participation of the states and tribes in planning and conducting the assessment within their respective jurisdictions. Not only are states and tribes key to survey implementation, they are the entities responsible under the Clean Water Act (CWA), Section 305b, to report to Congress regarding the extent to which the nation's waters support the CWA goals. From the research and development phase to the current operational program, numerous workshops and training sessions have been held to build technical expertise regarding monitoring design, sampling, data analysis, and interpretation of results. Through this technical transfer, numerous organizations have modified their local monitoring efforts to incorporate NCCA methods and approaches to assessing the condition of coastal resources. State and tribal partners have been active participants in the ongoing assessments of the performance of current indicators, and in the selection and testing of developmental indicators needed to respond to emerging environmental issues.

2.3 Assessing status and trends

Following the lead of earlier EPA coastal surveys, the NCCA approach had two primary goals regarding assessment: (1) evaluate the *status* of four major components of coastal ecosystems—the water column, sediment, and benthic and fish communities, and (2) ascertain how conditions *change over time (i.e., trends)*. For each of the key assessment components, conditions were evaluated based on a suite of core indicators and indices constructed from them. For instance, water quality was assessed using five measured indicators (concentrations of nutrients, chlorophyll, and dissolved oxygen, and water clarity) and a water quality index was then crafted from the five components. The assessment process first evaluated conditions at each site, rating each indicator and index as good, fair, or poor based on regionally determined assessment thresholds. Details regarding the indicators, indices, and thresholds used in assessments are presented in Section 3 of this chapter.

Once sites were evaluated, regional and national conditions were calculated. Recall that the survey design process had assigned each site a weighting factor equal to the area represented by the station. Regional assessments were then expressed as the percent of the region in good, fair, poor, or unassessed condition. For instance, the percent area of the Pacific coast in good condition was simply calculated as the sum of weighting factors associated with Pacific sites rated as good, divided by the total area of the Pacific coastal region (sum of all Pacific site weights). Assessments were calculated for the nation, for the five primary reporting regions (**Figure 2**), and for any state or designated research area containing a statistically-sufficient number of sites.

The survey design procedure further provided a measure of the uncertainty in the condition estimates, expressed as the 95th percentile confidence interval (CI), which was calculated as the binomial proportion confidence interval adjusted for possible spatial gradients in indicator measurements [23, 27]. Operationally, the confidence intervals were calculated using a complex computer-intensive algorithm, coded in the R-programing language, available at EPA's Aquatic Resource Monitoring website [25].

As the number of surveys conducted increases, the NCCA documents change over time. Typically, trends have been evaluated by analyzing what happens at an individual location, much as a physician monitors trends in the weight of an individual patient. In contrast, trends for NCCA were evaluated at the population level, i.e., trends in the proportion of sites in good condition. These population level trends were evaluated by noting statistically significant changes, i.e., condition estimates displaying non-overlapping CIs, determined over a series of comparable surveys. Since the early 1990s, coastal survey methods have evolved significantly over time. In some cases, new analyses can be applied to old data. In other cases, methodological differences have precluded trend analyses over the entire 30-year period. Eventually, trends in national assessments will reflect only NCCA surveys conducted from 2010 onward.

2.4 Communicating results

The approaches used to communicate survey results in summary reports reflect how the coastal survey approach evolved and innovated. Early regional EMAP surveys were essentially data reports prepared for a technical audience of monitoring practitioners. These terse reports emphasized methodology and reported results in tables, weighted-CDF plots, and bar plots e.g., [11]. While invaluable to technical staff and managers, these statistical summary reports attracted little public attention. In contrast, the national reports summarizing the EMAP-NCA surveys—the National Coastal Condition Reports NCCR I–IV [16–19]—were primarily prepared to be informative and understandable to the general public. These attractive and sizable documents were organized by region, featured highlights about local issues and showcased abundant photos and illustrations, as well as were available in hardcopy. In particular, NCCR-II and NCCR-III presented maps with site conditions portrayed by color-coded symbols. The NCCR reports' use of pie charts conveyed assessment results concisely and intuitively, but without adequate expression of uncertainty.

Beginning with the NARS-NCCA 2010, the reporting strategy changed substantially to accommodate the approach of conducting relatively standardized assessments on a regular schedule. The reports focused on delivering assessment results

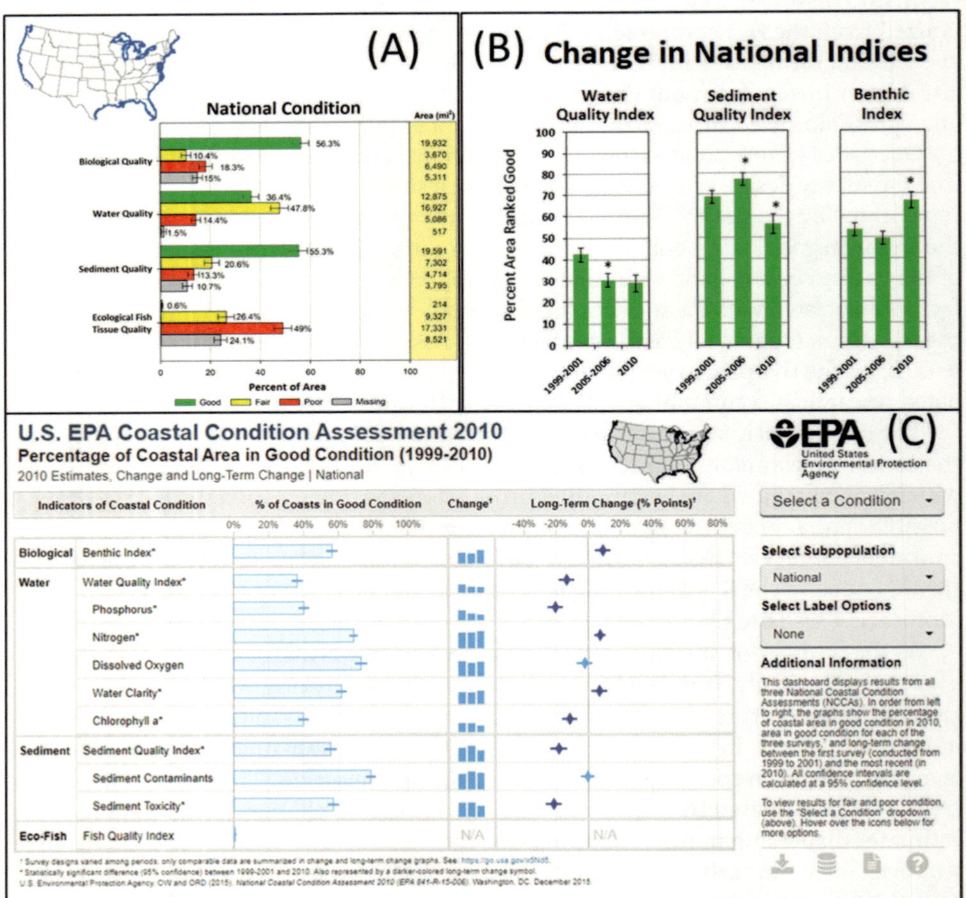

Figure 3.
Examples of coastal survey summary graphics from NCCA national reports highlighting national status in 2010 (A), trends 1999 to 2010 (B), and "dashboard" approach of reporting results (C).

concisely and quickly, primarily tailored for a technical audience of environmental managers. The reports are only accessible online and include fewer highlight-sections or explanatory graphics but continue to present material intuitively for public viewing. Graphics prominently display estimates of uncertainty and express change over time (**Figure 3**). The online 2015 NCCA report (in preparation) notably features an interactive "dashboard" graphic that allows the viewer to select the results in summary form as well as to access the data associated with the display. Importantly, the coastal reporting format is evolving in concert with the reporting approaches of other NARS surveys, thereby facilitating cross-resource assessment and modeling efforts.

3. NCCA method highlights

In this section we take a closer look at the methods used to assess the major components of coastal ecosystems—the water column, sediment, and benthic and fish communities. One issue was recognized early in the NCA program when national-scale surveys were undertaken—the U.S. coastal regions are extraordinarily diverse. The northeastern states reflect relatively late deglaciation, featuring minimal run-off from small watersheds into well-mixed coastal waters. Large drowned-river estuaries dominate the mid-Atlantic states, where environmental conditions are heavily influenced by the densely populated coastal communities. Estuaries along the southeastern states and the Gulf of Mexico reflect interaction with large, flat watersheds; these regions are subject to distinct sub-tropical bio-physical processes. In contrast, there are far fewer estuaries along the Pacific coast because of the absence of a coastal plain, and coastal processes there are uniquely affected by strong ocean currents and upwelling of cold, nutrient-rich water. How should surveys account for such diversity and differentiate natural from anthropogenic sources and responses?

In response to these challenges, survey planners initially relied on the advice of regional estuarine experts convened to suggest assessment indicators and provide benchmark values used to distinguish good, fair, and poor conditions. In reports we emphasized that these cut-points were appropriate for the surveys only, and generally distinct from regulatory thresholds. For each component assessment, several indicators of condition were evaluated separately and then combined into an overall index. In some cases, as is described below, the initial suite of indicators, indices, and benchmark values were modified and refined based on lessons learned. For instance, local benthic indices were replaced with a single index applicable nationwide; the fish community index was refashioned to better reflect ecological rather than human health conditions; and several human-health indicators were introduced. In the following sections, we describe the indicators and thresholds currently specifically employed in the NCCA surveys while highlight lessons learned from 30 years of experimenting and refining techniques.

3.1 Assessing water quality

The water column is a notoriously dynamic environment. Physical and biological process interact to create rapid and highly localized interactions of light, nutrients, algal growth and predation, and a host of quickly changing abiotic factors. Despite these challenges, deepening concerns regarding cultural eutrophication in coastal waters motivated survey planners to devise a strategy for assessing coastal water quality. Cultural eutrophication is the detrimental degradation of water quality often associated with nutrient over-enrichment [28, 29]. The NCCA assessment approach consisted of employing indicators that measure eutrophication-related symptoms and problems such as nutrient over-enrichment, excessive algal blooms, hypoxia or anoxia, low water transparency, etc. To moderate the inherent variability of such measures, the indicators were then combined into an index that is less dynamic than the individual components.

Table 2 lists the five core indicators and thresholds used in recent NCCA surveys to assess water quality in estuaries and the Great Lakes. Nutrient and chlorophyll concentrations were measured in surface water, dissolved oxygen levels were determined in bottom water, and water clarity was established at each site. These measures were then combined into a water quality index (WQI) that captured conditions likely to be indicative of problematic eutrophication regardless of when in summer sampling occurred [17]. For instance, the WQI might record excessive dissolved nutrients in early season, excessive algal production and poor water clarity in mid-season or hypoxic, turbid conditions in late season. Essentially, the WQI reflects a "preponderance of evidence"; the index is a more robust indicator of problematic eutrophication symptoms than the core indicators.

Thresholds generally varied by region. For instance, less stringent nutrient thresholds were specified for the West coast region, which experiences natural upwelling in summer [30], and more conservative guidelines were applied when assessing the tropical waters of southern Florida to protect submerged aquatic vegetation (SAV) beds. Assessment methods differed slightly when evaluating nutrients and water clarity in estuaries and the Great Lakes, recognizing the distinct ecologies and assessment histories in these environments. See details of the coastal water quality approach at pp. 11–15 of reference [19].

The NCCA approach of assessing nutrient status in estuaries continues to evolve. Early surveys measured nutrients as dissolved inorganic nitrogen and phosphorus (DIN and DIP). While DIN and DIP concentrations are valid indicators of nutrient enrichment status, they are unreliable measures of nutrient availability later in the season because they are generally assimilated into algal biomass in spring and early summer [19]. This is particularly problematic for NCCA surveys, which sample throughout the summer index period. In contrast, total nitrogen and phosphorus (TN and TP) are less variable and are related to chlorophyll concentrations [31]. Consequently, TN and TP were added as core indicators beginning in 2010 and were used to evaluate nutrient status in subsequent NCCA surveys. Since regional TN and TP thresholds have not yet been established, TN and TP were treated as exploratory indicators, rated as low, moderate, high, and very high based on the 25th, 50th, and 75th quartile values of the measured 2010 TN and TP values. The water quality index (WQI) continued to be calculated with DIN and DIP as described in **Table 2**, reflecting the key role of dissolved nutrients in eutrophication processes [17, 31].

3.2 Assessing sediment quality

Contaminants from agricultural, industrial, and nonpoint sources find their way to coastal waters where they may adsorb onto suspended particles and settle to the sediment. There, metals and organic pollutants are ingested by benthic-dwelling organisms and may become concentrated throughout the food web and adversely affect fish, pelagic mammals, and human consumers of aquatic organisms. To monitor sediment contamination, all EPA coastal assessments since the 1990s followed the approach of NOAA's Status and Trends program [32] and collected sediment grab samples and measured a suite of 74 metal, PAH, PCB, and pesticide contaminants in surficial sediment samples (**Table 1**). The impacts of the pollutants on benthic organisms were evaluated against the effects-based sediment quality guidelines, ERL (effects range low) and ERM (effects range median) [33].

Water quality indicators	Estuary thresholds	Great Lakes thresholds
Nitrogen status DIN & TN in surface water DIN used to assess N status in estuaries prior to 2015; TN used thereafter. Water Quality Index (WQI) constructed with DIN in estuaries; with TN in the GL	*DIN Thresholds (mgN/L): Good/Fair, Fair/Poor* NE/SE/Gulf: 0.1, 0.5 West: 0.35, 0.5 Tropical: 0.05, 0.1 *2015 TN Interim Thresholds (mgN/L):* 25th, 50th, 75th percentiles of NCCA 2010 TN values	Nitrogen not assessed in the Great Lakes
Phosphorus status DIP & TP in surface water DIP used to assess P status in estuaries prior to 2015; TP used thereafter. Water Quality Index (WQI) constructed with DIP in estuaries; with TP in the GL	*DIP Thresholds (mgP/L): Good/Fair, Fair/Poor* NE/SE/Gulf: 0.01, 0.05 West: 0.07, 1.0 Tropical: 0.005, 0.01 *2015 TP Interim Thresholds (mgN/L):* 25th, 50th, 75th percentiles of NCCA 2010 TP values.	*TP Thresholds (mgP/L): Good/Fair, Fair/Poor*/Superior/ Huron: 0.005, 0.01 Michigan: 0.007, 0.01 Saginaw/West Erie: 0.015, 0.032 Mid & East Erie/Ontario: 0.01, 0.015
Algal biomass Chlorophyll *a* in surface water	*Chla Thresholds (μg/L): Good/Fair, Fair/Poor* NE/SE/Gulf/West: 5.0, 20.0 Tropical: 0.5, 1.0	*Chla Thresholds (μg/L): Good/Fair, Fair/Poor*/Superior/ Huron: 1.3, 2.6 Michigan: 1.8, 2.6 Saginaw/Erie West: 3.6, 6.0 Mid & East Erie/Ontario: 2.6, 3.6
Oxygen status Dissolved oxygen (DO) in bottom water	*DO Thresholds (mg/L): Good/Fair, Fair/Poor* All coastal regions: 5.0, 2.0	*DO Thresholds (mg/L): Good/Fair, Fair/Poor* All lakes and basins: 5.0, 2.0
Water clarity *Estuaries*: Rated by fraction of PAR transmitted through 1 m. Thresholds vary by turbidity category *Great Lakes*: Rated by Secchi depth	*Transmissivity* @ 1m(%): Good/Fair, Fair/Poor* Naturally turbid waters: 10%, 5% Normally turbid waters: 20%, 10% SAV restoration priority: 40%, 20%	*Secchi Thresholds* (m): *Good/Fair, Fair/Poor*/Superior/Huron: 8.0, 5.3 Michigan: 6.7, 5.3 Saginaw/West Erie: 3.9, 2.1 Mid & East Erie/Ontario: 5.3, 3.9
Water Quality Index (WQI) Constructed based on the ratings of the measured component WQ metrics (five metrics in estuaries, including DIN & DIP; four metrics in the Great Lakes)	*Thresholds* **Good**: a maximum of one metric is rated as fair, and no metrics are rated as poor **Fair**: one metric is rated as poor, or two are rated as fair **Poor**: two or more metrics are rated as poor **Missing**: two metrics are missing, and available metrics do not suggest fair or poor ratings	*Thresholds* **Good**: a maximum of one metric is rated as fair, and no metrics are rated as poor **Fair**: one metric is rated as poor, or two are rated as fair **Poor**: two or more metrics are rated as poor **Missing**: two metrics are missing, and available metrics do not suggest fair or poor ratings

**Trans = exp(−Kd); where Kd is PAR extinction factor, calculated via regression of exponential attenuation of PAR intensity Iz/Io vs depth z, i.e., Beer's law: Iz/Io = exp(−Kd*z).*

Table 2.
Indicators and thresholds employed in the NCCA 2010 to assess water quality in estuaries and the Great Lakes.

ERL values are the concentration levels below which adverse bioeffects are unlikely, and ERM values signify the concentration above which adverse effects are likely. Sediments were also characterized by measuring grain size and total organic carbon (TOC) concentrations and were further tested for toxicity arising from either natural or anthropogenic sources by exposing amphipods to sediments in laboratory assays [34, 35].

Prior to the NCCA 2010, estuarine surveys evaluated sediment quality based on three core metrics: (1) sediment contaminants were evaluated as good, fair, or poor based on the number of ERL or ERM exceedances evident at a site; (2) toxicity was rated as good or poor if the survival rate of the amphipod *Ampelisca abdita* exceeded or was less than 80%, respectively; and (3) TOC was rated against concentration thresholds of 2 and 5%. A sediment quality index (SQI) was then calculated reflecting the ratings of the individual core components. Details are further explained in the National Coastal Condition Report IV [19].

Several modifications were introduced into NCCA surveys conducted in 2010 and later. Pollutant levels in estuaries were expressed as mean ERM quotients (mERMQ)—the ratio of a contaminant concentration to its ERM value, designated as mERMQ [36, 37]. Estuarine sediment contaminants were evaluated in a more nuanced manner, using the mERMQ and a logistic regression model approach [38] to better estimate the adverse effects of pollutants on benthic organisms. Estuarine sediment toxicity tests were primarily conducted using the amphipod species

Sediment quality indicators	Estuary thresholds	Great Lakes thresholds
Sediment contamination **Mean contaminant quotients**[a] For estuaries: calculate ERM-Q and mean ERM-Q (mERM-Q); For Great Lakes: calculate PEC-Q and mean PEC-Q (mPEC-Q). **Logistic regression model (LRM)**[b] For estuaries only: (1) calculate LRM factor for each of 36 analytes with fitting parameters; (2) select largest factor LRM_{max}; (3) calculate LRM P_{max}	**Good:** mERM-Q ≤ 0.1 & LRM $P_{max} \leq 0.05$ **Fair:** mERM-Q > 0.1 & ≤ 0.5 or LRM $P_{max} > 0.5$ & ≤ 0.75 **Poor:** mERM-Q > 0.5 or LRM $P_{max} > 0.5$	**Good:** mPEC-Q ≤ 0.1 **Fair:** mPEC-Q > 0.1 and ≤ 0.6 **Poor:** mean PEC-Q > 0.6
Sediment toxicity % Survival of amphipods after 10-day exposure to site sediment, compared with survival in clean control sediment Amphipods tested for estuarine sediments: *Leptocheirus plumulosus* or *Eohaustorius estuarius*; for Great Lakes sediments: *Hyalella azteca*	**Good**: test results not significantly different from control (p > 0.05) and ≥ 80% control-corrected survival **Fair**: test results significantly different from control (p ≤ 0.05) and ≥ 80% control-corrected survival or Test not significantly different from control (p > 0.05) and < 80% control-corrected survival **Poor**: test results significantly different from control (p < 0.05) and <80% control-corrected survival	**Good**: control corrected survival ≥90% **Fair**: control corrected survival ≥75 and <90% **Poor**: ontrol corrected survival <75%
Sediment Quality Index (SQI) Constructed based on the ratings of sediment contaminant and sediment toxicity metrics The assessment criteria are the same for estuarine and Great Lakes sites	**Good**: both sediment contaminant and sediment toxicity metrics are rated good **Fair**: neither metric is rated poor and at least one metric is rated fair **Poor**: at least one metric is rated poor	**Same assessment criteria**

[a]*ERM-Q, conc/ERM for estuarine sites only, for 28 analytes with ERM values; PEC-Q, conc/PEC for Great Lakes only, for 9 analytes with PEC values (As, Cd, Cr, Cu, Pb, Ni, Zn, PAHs, PCBs); mean ERM-Q, ΣERM-Q/n; mean PEC-Q = ΣPEC-Q/n; where n = number or analytes.*
[b]*LRM factor calculated as follows, for 36 analytes with fitting factors B_0 and B_1 (Field et al., 2002).*
$LRM\ P_{max} = 0.11 + 0.33*LRM_{max} + 0.4*LRM_{max}^2$.

Table 3.
Indicators and thresholds employed in the NCCA 2010 to assess sediment quality in estuaries and the Great Lakes.

Leptocheirus plumulosus and *Eohaustorius estuarius* (in California), which could be cultured in the laboratory and gave more consistent results than *Ampelisca abdita*. TOC was no longer used to assess sediment quality because it could have positive and negative impacts on organisms, complicating interpretation. The sediment quality index (SQI) was constructed from the remaining two core metrics to summarize overall sediment condition. Similar to the estuarine approach, the Great Lakes sediment quality index utilizes a mean sediment quality guideline quotient method and a toxicity test. The mean Probable Effect Concentration Quotient (mPEC-Q), rather than the mERM-Q, is used in the Great Lakes [39], along with using the freshwater amphipod *Hyalella azteca* to assess toxicity (**Table 3**). Further details concerning the evolution and calculation of methods are available in the NCCA 2010 technical report [40].

3.3 Assessing benthic community condition

All EPA estuarine surveys since the 1990s collected sediment grab samples of benthic macroinvertebrate communities for assessment of ecological condition based on measures of diversity, species richness, and dominance. The benthos is a key component of estuarine ecosystems, serving as important food source for higher trophic levels and maintaining sediment and water quality. Benthic communities respond to contaminant concentrations, dissolved oxygen stress, salinity fluctuations and physical disturbance, and are relatively immobile and therefore integrate the effect of adverse conditions over months and years.

Separate regional, benthic-community condition indices were developed during the EMAP programs of the 1990s, including for the Virginian [41], Carolinian [42] and Louisianan [43, 44] biogeographic provinces. Later, an index was created for the Acadian Province (Maine through Cape Cod waters) [45]. No specific index was developed for the Pacific coast; rather, sites were assessed based on observed vs. expected species richness [30]. These benthic indices were used in estuarine assessments prior to NCCA 2015. Benthic communities in the Great Lakes were evaluated using an oligochaete trophic index (OTI) based on the classification of oligochaete species by their tolerance to organic enrichment [46, 47]. **Table 4** presents a summary of these regional benthic indices. The NCCA 2010 Technical Appendix [40] provides further detail regarding the development and calculation of the indices.

While the separate estuarine indices performed well in the region for which they were developed, they were developed using different statistical models and metrics. Because the different indices might not be comparable, combining the separate indices into a nationwide evaluation tool was problematic. In response, a national-scale index called M-AMBI (multivariate-AZTI marine biotic index) was adapted to provide a single index applicable to all U.S. estuarine waters [48, 49]. This index is based on benthic indices that were successfully deployed in Europe and elsewhere [50, 51]. AMBI is an abundance-weighted tolerance index, while M-AMBI combines AMBI, species richness and species diversity together using factor analysis calculated for a given habitat. The resulting index was shown to be comparable to several local indices [49] and was better correlated with land use variables [52]. The resulting scores are based on comparison of a sites' position along a pollution gradient [49].

3.4 Assessing fish tissue contaminants

Many aquatic organisms in coastal regions are inadvertent inheritors of a legacy of disturbances often associated with human practices. For instance, chemical pollutants from farms and cities delivered to coastal waters enter the food web

Region/province	Method	Component metrics	References
Northeast/ Acadian	Logistic regression analysis	Shannon H' (diversity) MN_ES(50)0.05 (species tolerance index) % Capitellid polychaetes (abundance)	[39]
Northeast/ Virginian	Discriminant analysis	Salinity adjusted Gleason D (diversity) Salinity adjusted % tubificid (abundance) % Spionids (abundance)	[35]
Southeast/ Carolinian	B-IBI approach	Mean abundance Mean number of taxa 100% abundance of 2 dominant taxa % Abundance of pollution sensitive taxa	[36]
Gulf/Louisianian	Discriminant analysis	% Expected diversity (Shannon H') Mean abundance of tubificids % Capitellids % Bivalves % Amphipods	[37, 38]
Pacific coast	Regression	Observed vs expected species richness	[11, 27]
Great Lakes	Abundance-weighted tolerance equation	Oligochaete tolerance scores (based on organic enrichment)	[40, 41]
National estuarine	Factor analysis	Shannon H' (diversity) Species richness AMBI (abundance-weighted pollution tolerance)	[42, 43]

The national M-AMBI index was developed for the NCCA 2015 and future surveys.

Table 4.
Summary of methods, metrics, and thresholds used to construct regional benthic indices used to evaluate assess coastal waters.

and accumulate, threatening fish and higher trophic-level communities, humans included. To assess the ecological danger to aquatic communities, EPA's coastal surveys since the 1990s have measured concentrations of metals, PCBs, PAHs, and pesticides (**Table 1**) in demersal and pelagic fish collected at sampling stations. Prior to the NCCA 2010 survey, sites were evaluated by comparing contaminant concentrations against human health fish-consumption advisory thresholds as a surrogate for ecologically-relevant benchmarks [53]. When both humans and wildlife were similarly sensitive to specific contaminant exposures, the surrogate used for the assessment was meaningful. Beginning with the NCCA 2010 survey, an *ecological* risk-based approach using wildlife endpoints was incorporated to better align with the ecosystem focus of the NCCA surveys.

The ecological risk approach assessed contaminant levels in whole-body fish tissue following the methods of EPA's ecological risk assessment [54]. The primary goal of this NCCA index, therefore, was to evaluate the potential risk that consuming contaminated fish poses to predators other than humans. Because such "wild" predators consume the entire fish, the NCCA protocol measured contaminant concentrations in the entire fish collected in the survey, rather than measuring contaminant levels in just the fillet—the protocol formerly used when human health was the focus. Operationally, the process first identified mammalian, avian, and piscivorous "receptors," i.e., predator species that consume coastal fish and could be adversely affected by contaminants in the prey-fish. **Table 5** lists the freshwater and marine receptors selected for analysis based on their diet (predominantly fish) and availability of data in the literature. The literature studies were reviewed to identify the Lowest Observed Adverse Effects Level or LOAEL for each receptor, that is, the contaminant concentration likely to elicit toxicological effects. The minimum contaminant LOAEL found for any member of a receptor group was designated as an impairment threshold, and was used to rate survey sites as good, fair, or poor (**Table 6**). Because of the very different methods used in the human-health and ecological-risk approaches, the NCCA assessments cannot be directly compared

Avian receptors	Freshwater mammalian receptors	Marine mammalian receptors	Freshwater fish receptors	Marine fish receptors
Great Blue Heron	River Otter	Harbor Seal	Largemouth Bass	Bluefin Tuna
Osprey	Mink	Bottlenose Dolphin	Florida Gar	Yellowfin Tuna
Bald Eagle	—	Walrus	Muskellunge	Shortfin Mako
Herring Gull	—	—	Snakehead	Mackerel Tuna
Belted Kingfisher	—	—	Lake Walleye	Swordfish
Brown Pelican	—	—	—	—

Table 5.
Higher trophic-level piscivores potentially at risk from consuming contaminated prey fish.

Contaminant	Whole-body tissue concentration (μg/dry g) by receptor group		
	Lowest observed adverse effect level (LOAEL)		
	Mammal	Avian	Fish
Arsenic (inorganic)	3.8	9.2	0.7
Cadmium	32.1	14.0	3828
Mercury (methyl)	1.1	0.1	1.4
Selenium	2.3	0.6	33.6
Chlordane	55.4	2.9	—
DDTs	28.0	1.6	7.1
Dieldrin	1.2	0.3	1.6
Endosulfan	42.8	43.2	0.003
Endrin	5.6	0.1	3.9
Heptachlor epoxide	7.5	6.3	81.1
Hexachlorobenzene	14.0	0.6	0.04
Lindane	280	2.4	376
Mirex	4.6	0.7	9.9
Toxaphene	280	3.6	0.03
PCBs	3.9	1.3	2.0
Rating criteria for ecological fish tissue contaminant index			
Good	**Fair**		**Poor**
No contaminant concentration exceeds a LOAEL for any receptor group	At least one contaminant concentration exceeds a LOAEL for one receptor group		At least one contaminant concentration exceeds a LOAEL for two or more receptor groups

Table 6.
Whole-body tissue contaminant LOAEL concentrations (μg/dry g) by receptor group.

with earlier survey results and cannot be used to inform human consumption advisories. Refer to the NCCA 2010 technical appendix for further details [34].

3.5 Addressing human-health concerns and emerging issues

Along with evaluating the ecological condition of several major ecological compartments of coastal ecosystems, the NCCA also addressed several matters regarding human health and emerging issues. For instance, in the Great Lakes with the support of the Great Lakes Restoration Initiative (GLRI) [55], the concentrations in fish tissue of the contaminants mercury, polychlorinated biphenyls (PCBs), flame retardant polybrominated diphenyl ethers (PBDEs), and perfluorinated compounds (PFCs) were measured in the Great Lakes NCCA surveys and evaluated against human health screening values [40]. The NCCA also initiated a survey-wide monitoring program quantifying aqueous concentrations of the algal toxins microcystin and cylindrospermopsin, as well as mercury in fish muscle. Several exploratory studies were also undertaken to address important issues such as ocean acidification and the distribution of micro-plastics in coastal water. Newer assessment techniques are also under investigation, such as exploring the use of underwater cameras and environmental genetic screening to monitor the expansion of invasive organisms in the Great Lakes.

4. Conclusion

In retrospect, the mandate issued to the U.S. EPA by the Clean Water Act in 1972 to compile a national assessment of water quality was a bold and challenging directive. No blueprint was available to indicate the best approach of conducting a large-scale assessment program. Tactics regarding monitoring designs, sampling strategies, indicators, thresholds, assessment protocols, etc. all needed to be developed from scratch. The EPA adopted a pragmatic approach to assessing coastal regions, exploring and testing methodologies regionally, and then gradually building a national program based on the best practices learned over 30 years of experimentation. While the NARS-coastal surveys and assessments are not perfect, they represent the first nationally consistent effort, based on current practices, to assess the Nation's coastal waters through time. The data and results represent information available for evaluating national policy and a basis for the scientific community to evaluate coastal waters from many perspectives.

The evolution of methodologies and approaches for the NCCA is an ongoing process. Future surveys will continue the practices of adapting current methods to the latest best practices and the adaptation of new strategies, while striving to strike a balance between consistency and creative exploration. The continued importance of partnerships among federal, state and tribal agencies cannot be over-emphasized in achieving the aims of the monitoring program. Such cooperation has proven to be both efficient and productive, and the enhanced capacity of states and tribes to conduct assessments independently is particularly valuable in assuring a sustainable monitoring program. Particularly striking has been the deep commitment of many individuals, research scientists, program planners, crew members, information managers, analysts, communicators, and partners, who have offered feedback and criticism to continuously improve the coastal assessment process. Finally, the development and evolution of coastal assessment expertise described in this chapter is similarly evident in sister NARS programs that assess lakes, rivers and streams, and wetlands. Descriptions of these programs are presented elsewhere in this book.

Acknowledgements

These coastal assessments were conducted in partnership with many state and tribal agencies, the National Oceanic and Administration (NOAA), the National Park Service (NPS), as well as other federal agencies. EPA offices included the Office of Research and Development, the Office of Water, Great Lakes National Program Office, and EPA Regional Monitoring Coordinators from Regions 1, 2, 3, 4, 5, 6, 9, and 10. We note the particular efforts of EPA colleagues who contributed to the conception, administration and operation of the coastal surveys, including Kevin Summers, John Paul, Virginia Hansen, Treda Grayson, Sarah Lehmann, and Greg Colianni and many others too numerous to name. Thanks also to EPA colleagues Don Cobb, Charlie Strobel, and Hal Walker who reviewed the chapter. We especially wish to acknowledge the truly invaluable assistance of the hundreds of participants who conducted the assessments, including the field crews, biologists, taxonomists, statisticians, analysts, program administrators, regional coordinators, project managers, quality control officers, and reviewers.

Conflict of interest

The authors declare no conflict of interest.

Disclaimer

The views expressed in this chapter are those of the authors and do not necessarily represent the views or policies of the United States Environmental Protection Agency.

Author details

John A. Kiddon[1], Hugh Sullivan[2]*, Walter G. Nelson[3], Marguerite C. Pelletier[1], Linda Harwell[4], Mari Nord[5] and Steve Paulsen[6]

1 U.S. Environmental Protection Agency, Office of Research and Development, Center for Environmental Measurement and Modeling, Atlantic Coastal Environmental Sciences Division, Narragansett, RI, USA

2 U.S. Environmental Protection Agency, Office of Water, Office of Wetlands, Oceans and Watersheds, Washington, DC, USA

3 U.S. Environmental Protection Agency, Office of Research and Development, Center for Public Health and Environmental Assessment, Pacific Ecological Systems Division, Newport, OR, USA

4 U.S. Environmental Protection Agency, Office of Research and Development, Center for Environmental Measurement and Modeling, Gulf Ecosystem Measurement and Modeling Division, Gulf Breeze, FL, USA

5 U.S. Environmental Protection Agency, Region 5, Water Division, Chicago, IL, USA

6 U.S. Environmental Protection Agency, Office of Research and Development, Center for Public Health and Environmental Assessment, Pacific Ecological Systems Division, Corvallis, OR, USA

*Address all correspondence to: sullivan.hugh@epa.gov

References

[1] U.S. EPA. Summary of the Clean Water Act. 2019. Available from: https://www.epa.gov/laws-regulations/summary-clean-water-act; and https://www.epa.gov/sites/production/files/2017-08/documents/federal-water-pollution-control-act-508full.pdf

[2] U.S. General Accounting Office. Better Monitoring Techniques Are Needed to Assess the Quality of Rivers and Streams. Volume 1. CED-81-30. Washington, DC: U.S. General Accounting Office; 1981

[3] National Research Council. Environmental Monitoring. Washington, DC: Volume IV. National Academy of Sciences; 1977

[4] U.S. Environmental Protection Agency. Surface Water Monitoring: A Framework for Change. Washington, DC: Water Office; 1987

[5] U.S. Environmental Protection Agency. Report of the National Workshop on Instream Biological Monitoring and Criteria. Corvallis, OR: Environmental Research Laboratory-Corvallis; 1987

[6] U.S. General Accounting Office. Protecting human health and the environment through improved management. GAO/RCED-88-101. Gaithersburg, MD: U.S. General Accounting Office; 1988

[7] U.S. Environmental Protection Agency. Framework for ecological risk assessment. EPA/630/R-92/001. Washington, DC: Risk Assessment Forum, USEPA; 1992

[8] Messer JJ, Linthurst RA, Overton WS. An EPA program for monitoring ecological status and trends. Environmental Monitoring and Assessment. 1991;**17**:67-78

[9] Holland AF. Near Coastal Program Plan for 1990: Estuaries. Washington, DC: Office of Research and Development; 1990. EPA 600/4-90/033

[10] Summers JK, Paul JF, Robertson A. Monitoring the ecological condition of estuaries in the United States. Toxicological and Environmental Chemistry. 1995;**49**:93-108

[11] Nelson WG, Lee H, Lamberson JO, Engle V, Harwell L, Smith LM. Condition of Estuaries of Western United States for 1999: A Statistical Summary. Washington, DC: Office of Research and Development, National Health and Environmental Effects Research Laboratory; 2004. EPA/620/R-04/200

[12] Hyland JL, Snoots TR, Balthis WL. Sediment quality of estuaries in the southeastern U.S. Environmental Monitoring and Assessment. 1998;**51**:331-343

[13] Paul JF, Gentile JH, Scott KJ, Schimmel SC, Campbell DE, Latimer RW. EMAP-Virginian Province four-year assessment (1990-93). EPA/620/R-99/004. Washington, DC: Office of Research and Development; 1999

[14] Kiddon JA, Paul JF, Buffum HW, Strobel CS, Hale SS, Cobb D, et al. Ecological condition of U.S. Mid-Atlantic estuaries, 1997-1998. Marine Pollution Bulletin. 2003;**46**:1224-1244

[15] Summers JK, Engle VD, Smith LM, Harwell LM, Macauley JM, Harvey JE. Monitoring the condition of the estuaries in the United States. In: Jordan S, editor. Estuaries: Classification, Ecology, and Human Impacts. Hauppauge, NY: Nova Science Publishers, Inc; 2012. pp. 249-268

[16] National Coastal Condition Report I. EPA-620-R-01-005. 2001.

Washington, DC: U.S. Environmental Protection Agency, Office of Research and Development and Office of Water. Available from: https://archive.epa.gov/nheerl/arm/web/html/monit_intro.html

[17] National Coastal Condition Report II. EPA-620-R-03-003. 2004. Washington, DC: U.S. Environmental Protection Agency, Office of Research and Development and Office of Water. Available from: https://archive.epa.gov/nheerl/arm/web/html/monit_intro.htm

[18] National Coastal Condition Report III. EPA-842R-08-002. 2008. Washington, DC: U.S. Environmental Protection Agency, Office of Research and Development and Office of Water, Available from: https://archive.epa.gov/nheerl/arm/web/html/monit_intro.html

[19] National Coastal Condition Report IV. EPA-842-R-08-003. 2010. Washington, DC: U.S. Environmental Protection Agency, Office of Research and Development and Office of Water. Available from: https://archive.epa.gov/nheerl/arm/web/html/monit_intro.html

[20] National Coastal Condition Report 2010. EPA-841-R-15-006. 2015. Washington, DC: U.S. Environmental Protection Agency, Office of Research and Development and Office of Water. Available from: https://www.epa.gov/national-aquatic-resource-surveys/ncca

[21] National Aquatic Resources Surveys (NARS). Available from: https://www.epa.gov/national-aquatic-resource-surveys

[22] National Coastal Condition Assessment. Basics of the NCCA/Manuals. Available from: https://www.epa.gov/national-aquatic-resource-surveys/ncca

[23] Olsen AR, Stevens DL Jr, White D. Application of global grids in environmental sampling. Computing Science and Statistics. 1998;**30**:279-284

[24] Urquhart NS, Paulsen SG, Larsen DP. Monitoring for policy-relevant regional trends over time. Ecological Applications. 1998;**8**:246-257

[25] Aquatic Resources Monitoring (USEPA). Available from: https://archive.epa.gov/nheerl/arm/web/html/monit_intro.html

[26] Data from the National Aquatic Resources Surveys. Available from: https://www.epa.gov/national-aquatic-resource-surveys/data-national-aquatic-resource-surveys

[27] Stevens DL, Olsen AR. Variance estimation for spatially balanced samples of environmental resources. Environmetrics. 2003;**14**:593-610

[28] Paerl HW, Hall NS, Peierls BL, Rossignol KL. Estuarine and coastal eutrophication dynamics in a culturally and climatically stressed world. Estuaries and Coasts. 2014;**37**:243-258

[29] Bricker S, Longstaff B, Dennison W, Jones A, Boicourt K, Wicks C, et al. Effects of nutrient enrichment in the Nation's estuaries: A decade of change. In: NOAA Technical Memorandum NOS NCCOS Coastal Ocean Program Decision Analysis Series 26. Silver Spring, MD: National Center for Coastal Ocean Science; 2007. p. 328

[30] Nelson WG, Brown CA. Use of probability-based sampling of water-quality indicators in supporting development of quality criteria. ICES Journal of Marine Science. 2008;**65**:1421-1427

[31] Smith VH. Responses of estuarine and coastal marine phytoplankton to nitrogen and phosphorus enrichment. Limnology and Oceanography. 2006;**51**:377-384

[32] O'Connor TP, Ehler CN. Results from the NOAA National Status and trends program on distribution and

effects of chemical contamination in the coastal and estuarine United States. Environmental Monitoring and Assessment. 1991;**17**:33-41

[33] Long ER, MacDonald DD, Smith SL, Calder FD. Incidence of adverse biological effects within ranges of chemical concentrations in marine and estuarine sediments. 1995. Environmental Management. 1995;**19**:81-97

[34] ASTM. Guide for conducting 10-day static sediment toxicity tests with marine and estuarine amphipods. In: American Testing and Materials ASTM Standard Methods. Section 11.04, Methods Number E-1367-90. Philadelphia, PA: American Society for Testing and Materials;

[35] Strobel C, Klemm DJ, Lobring L, Eichelberger J, Alford-Stevens A, Potter B, et al. EMAP-Estuaries Laboratory Methods Manual, Volume 1—Biological and Physical Analysis. EPA/620/R-95/008. Washington, DC: U.S. Environmental Protection Agency, Office of Research and Development. 1996

[36] Long ER, Field LJ, MacDonald DD. Predicting toxicity in marine sediments with numerical sediment quality guidelines. Environmental Toxicology and Chemistry. 1998;**17**:714-727

[37] Long ER, Ingersoll CG, MacDonald DD. Calculation and uses of mean sediment quality guideline quotients: A critical review. Environmental Science & Technology. 2006;**40**:1726-1736

[38] Field LJ, Macdonald DD, Norton SB, Ingersoll CG, Severn CG, Smorong D, et al. Predicting amphipod toxicity from sediment chemistry using logistic regression models. Environmental Toxicology and Chemistry. 2002;**21**:1993-2005

[39] MacDonald DD, Ingersoll CG, Berger TA. Development and evaluation of consensus-based sediment quality guidelines for freshwater ecosystems. Archives of Environmental Contamination and Toxicology. 2000;**39**:20-31

[40] National Coastal Condition Assessment (NCCA). 2010. Technical Report. Available from: https://www.epa.gov/sites/production/files/2016-01/documents/ncca_2010_technical_report_20160127.pdf

[41] Paul JF, Scott KJ, Campbell DE, Gentile JH, Strobel CS, Valente RM, et al. Developing and applying a benthic index of estuarine condition for the Virginian Biogeographic Province. Ecological Indicators. 2001;**1**:83-99

[42] Van Dolah RF, Hyland JL, Holland AF, Rosen JS, Snoots TR. A benthic index of biological integrity for assessing habitat quality in estuaries of the southeastern USA. Marine Environmental Research. 1999;**48**:269-283

[43] Engle VD, Summers K, Gaston GR. A benthic index of environmental condition of Gulf of Mexico estuaries. Estuaries. 1994;**17**:372-384

[44] Engle VD, Summers JK. Biogeography of benthic macroinvertebrates in estuaries along the Gulf of Mexico and western Atlantic coasts. Hydrobiologia. 2000;**436**:17-33

[45] Hale SS, Heltshe JF. Signals from the benthos: Development and evaluation of a benthic index for the nearshore Gulf of Maine. Ecological Indicators. 2008;**8**:338-350

[46] Howmiller RP, Scott MA. An environmental index based on relative abundance of oligochaete species. Journal of the Water Pollution Control Federation. 1997;**49**:809-815

[47] Environmental Canada and the U.S. Environmental Protection Agency. 2014. State of the Great Lakes 2011. Cat. No. En161-3/1-2011E-PDF. EPA 950-R-13-002. Available from: https://binational.net

[48] Gillett DJ, Weisberg SB, Grayson T, et al. Effect of ecological group classification schemes on performance of the AMBI benthic index in U.S. coastal waters. Ecological Indicators. 2015;**50**:99-107

[49] Pelletier MC, Gillett DJ, Hamilton A, Grayson T, Hansen V, Leppo EW, et al. Adaptation and application of multivariate AMBI (M-AMBI) in U.S. coastal waters. Ecological Indicators. 2018;**89**:818-827

[50] Borja A et al. A marine biotic index to establish the ecological quality of soft-bottom benthos within European Estuarine and Coastal Environments. Marine Pollution Bulletin. 2000;**40**(12):1100-1114

[51] Muxika I, Borja A, Bald J. Using historical data, expert judgement and multivariate analysis in assessing reference conditions and benthic ecological status, according to the European Water Framework Directive. Marine Pollution Bulletin. 2007;**55**:16-29

[52] Pelletier MC, Gold AJ, Copeland J, Gonzalez L, August PV. Landscape structure and land use affects estuarine benthic invertebrates in the Virginian Biogeographic Province, USA. Environmental Monitoring and Assessment. 2019;**191**:252

[53] U.S. Environmental Protection Agency. Guidance for Assessing Chemical Contaminant Data for Use in Fish Advisories, Volume 2: Risk Assessment and Fish Consumption Limits. Washington, DC: U.S. Environmental Protection Agency, Office of Water; 2000. EPA-823-B00-008

[54] U.S. Environmental Protection Agency. Ecological Risk Assessment Guidance for Superfund: Process for Designing and Conducting Ecological Risk Assessments. Edison, NJ: U.S. Environmental Protection Agency, Office of Solid Waste and Emergency Response, Environmental Response Team; 1997

[55] Great Lakes Restoration Initiative [Internet]. 2020. Available from: https://www.glri.us/ [Accessed: 03 March 2020]

Chapter 9

Wetland Assessment: Beyond the Traditional Water Quality Perspective

Mary E. Kentula, Amanda M. Nahlik, Steven G. Paulsen and Teresa K. Magee

Abstract

Use of water chemistry or water quality data as the sole indicator to determine if aquatic ecosystems meet restoration objectives or Clean Water Act criteria is not possible for wetland resources because surface water presence varies across wetland types. The 2011, National Wetland Condition Assessment (NWCA) assessed 967 sites representing 25,153,681 ha of wetland across the conterminous US. Surface water could be collected at 537 sites representing only 41% of the wetland population area and under-representing particular wetland types. These results motivated the authors to introduce the concept of *aquatic resource quality*, the condition of an ecosystem based on the integrated assessment of physical, chemical, and biological indicators, as the goal of monitoring and assessment of aquatic systems. The NWCA is an example of the use of *aquatic resource quality*. The survey successfully reported on wetland condition using a biotic indicator (the vegetation multimetric index) and the relative extent and relative risk of stressors using 10 physical, chemical, and biological indicators to report on *aquatic resource quality*. The NWCA demonstrated that *aquatic resource quality* can be consistently evaluated regardless of surface water presence. Consequently, we recommend *aquatic resource quality* as the goal of aquatic ecosystem monitoring and assessment.

Keywords: wetlands, monitoring and assessment, National Wetland Condition Assessment, aquatic resource quality, National Aquatic Resource Surveys, water chemistry, water quality

1. A new paradigm: *Aquatic resource quality*

For many, the terms water quality and water chemistry are synonymous, but others (e.g., Eriksson [1]) recognize a subtle yet important distinction between the terms. While the term water chemistry refers to the chemical composition of the water; water quality implies a value judgment on the suitability of the composition of the water for a specific use. Typically, the composition of the water is defined by chemical characteristics (as in Eriksson [1]), but sometimes physical or biological aspects of the water, such as turbidity, color, or odor, are used. Making a distinction between the definitions of water chemistry and water quality is essential for clear communication. To further avoid the ambiguities surrounding the use these terms, we introduce the concept of *aquatic resource quality* for reporting based on the physical, chemical, and biological integrity of aquatic resources as outlined in the goals of the Clean Water Act (CWA) [2].

Aquatic resource quality is defined herein as the condition of an aquatic ecosystem. Evaluating *aquatic resource quality* requires the integrated use of physical, chemical, and biological indicators to describe the condition of the resource and identify factors negatively affecting the condition [3]. Wetlands are an excellent test case for examining the application of the *aquatic resource quality* concept because traditional use of only water chemistry or water quality to determine whether rivers, streams, and lakes meet CWA criteria is not consistently possible for wetlands. Wetlands do not always have surface water. This is because the surface water in wetlands varies on seasonal and annual time scales, with regimes ranging from permanently flooded to saturated (i.e., substrate is saturated to the surface for extended periods, but surface water is seldom present) to intermittently flooded (i.e., weeks, months, or years may intervene between periods of inundation) [4]. Furthermore, certain wetland types, like fens, are groundwater-driven and rarely have surface water. Because sampling surface water for determination of chemistry is not always possible in wetlands, the adoption of the *aquatic resource quality* concept is required to holistically characterize the wetland resource.

Wetlands are a critical part of the Nation's aquatic resources and are protected under the CWA. Because of this, there is an obligation to include wetlands in monitoring programs for reporting under the CWA—despite the challenges associated with sampling wetlands. Fortunately, there is ample evidence that the wetland resource can be successfully assessed at large scales based on the *aquatic resource quality* concept (e.g., [5–7]). This early research helped inform the development of the National Wetland Condition Assessment (NWCA), which was first conducted in 2011 by the US Environmental Protection Agency (USEPA) to fulfill the objective of determining a baseline for *wetland resource quality* in the conterminous US. The goals of the NWCA were to:

- "produce a national report describing the condition of the Nation's wetlands and anthropogenic stressors commonly associated with poor condition;

- collaborate with states and tribes in developing complementary monitoring tools, analytical approaches, and data management technology to aid wetland protection and restoration programs; and

- advance the science of wetland monitoring and assessment to support wetland management needs" [8].

In this chapter, we present a summary of the 2011 NWCA design and methods, and then use national-scale data to report on patterns in the distribution of the wetlands represented by surface water chemistry. Finally, we examine how the NWCA fulfills the more comprehensive objective of reporting on *wetland resource quality* in accordance with the CWA requirements to consider the physical, chemical, and biological integrity of the wetland resource.

2. Data collection for the 2011 NWCA

The following subsections provide a brief overview of the design and field sampling methods used in the NWCA. For details see the 2011 NWCA Site Evaluation

Guidelines [9], Field Operations Manual [10], Laboratory Methods Manual [11], and Technical Report [12]. These documents are available on the NWCA website (https://www.epa.gov/national-aquatic-resource-surveys/nwca).

2.1 2011 NWCA survey design

The target population, that is, the specific portion of the wetlands of the conterminous United States (US) to be assessed in the 2011 NWCA was composed of tidal and nontidal wetlands with rooted vegetation and, when present, open water less than 1 m deep, and includes farmed wetlands not in crop production at the time of the survey [8]. The target population was comprised of seven of the wetland classes used in the US Fish and Wildlife Service's (USFWS) Wetlands Status and Trends (S&T) reporting [13]: Estuarine Intertidal Emergent (E2EM), Estuarine Intertidal Forested/Scrub Shrub (E2SS), Palustrine Emergent (PEM), Palustrine Farmed (Pf), Palustrine Forested (PFO), Palustrine Scrub Shrub (PSS), and Palustrine Unconsolidated Bottom/Aquatic Bed (PUBPAB). These classes are an adaptation of those defined by Cowardin et al. [4] and used in USFWS National Wetland Inventory (NWI) mapping.

A spatially balanced probability survey design [14–16] was developed using plots from the USFWS S&T Program as a basis for a sample of site locations for the NWCA. The USFWS S&T plots were mapped using 2005 aerial photography. The S&T Program mapped additional plots on the Pacific Coast at the request of the NWCA to assure sites would be selected for sampling along the coast due to the lower frequency of wetland occurrence in the Western US than in other parts of the country (**Figure 1**). The NWCA design allocated site locations by state and wetland class, generating 1800 potential site locations to ensure approximately 900 sites meeting target criteria would be available for sampling [12, 17]. Nine-hundred sites allow evaluation of different wetland types in the conterminous US and five major ecoregions. Ultimately, 967 sites from the probability design were sampled (**Figure 1**).

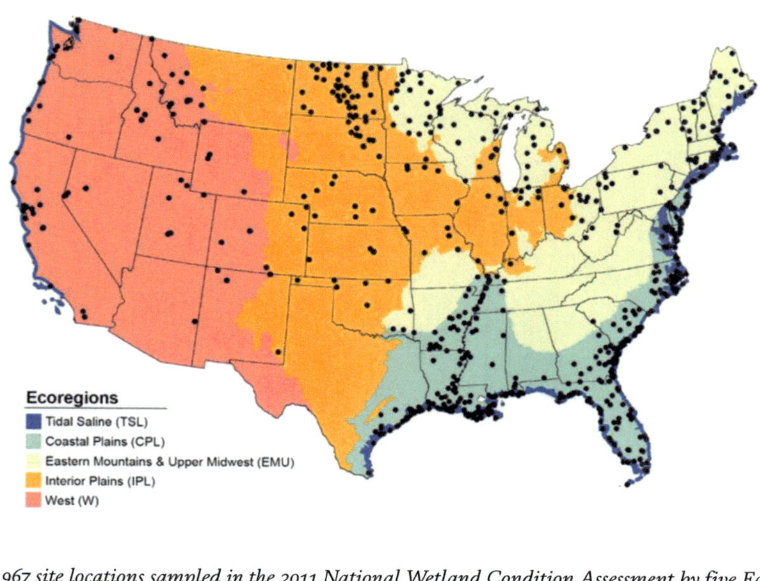

Figure 1.
Map of the 967 site locations sampled in the 2011 National Wetland Condition Assessment by five Ecoregions: Tidal Saline (TSL), Coastal Plains (CPL), Eastern Mountains & Upper Midwest (EMU), Interior Plains (IPL), and West (W). Note that CPL, EMU, IPL, and W exclusively include freshwater wetlands. The pattern of site locations reflects the distribution of wetlands across the conterminous United States with most wetland areas in the East and Southeast and the least in the Midwest and West.

As part of the design process, weights were assigned to each of the 1800 potential site locations that indicate the wetland area (i.e., the number of hectares) of the NWCA target population represented by the site (Olsen et al. [17]). After the 967 sites were visited, the weights were adjusted to account for the inability to sample sites, for example, due to denial of access, a site being inaccessible (i.e., safety issues), or a site failing to meet the target criteria (i.e., non-target). Finally, the adjusted weights were used to calculate the extent estimates of the wetland resource, expressed as hectares or percent of the wetland area, for different groupings (or subpopulations) of wetlands. The subpopulations presented in the 2011 NWCA final report (USEPA [8]) were ecoregion and wetland type. For a more detailed description of how this was done, see Diaz-Ramos et al. [18], Kincaid and Olsen [19], and Olsen et al. [17].

2.2 Field sampling for the 2011 NWCA

NWCA protocols for sampling each site were designed to be completed by a four-person field crew during a single day during peak growing season when most plants are in flower or fruit to optimize species identification and characterization of species abundance. This typically occurs between April and September depending on the status of the vegetation for sampling at the location of the site [10, 20]. The standard assessment area (AA) was a 0.5-ha circular plot with a 40-m radius, centered on the site location from the design (**Figure 2**). A buffer extended 100 m from the edge of the AA. If the wetland size and shape made the standard, circular

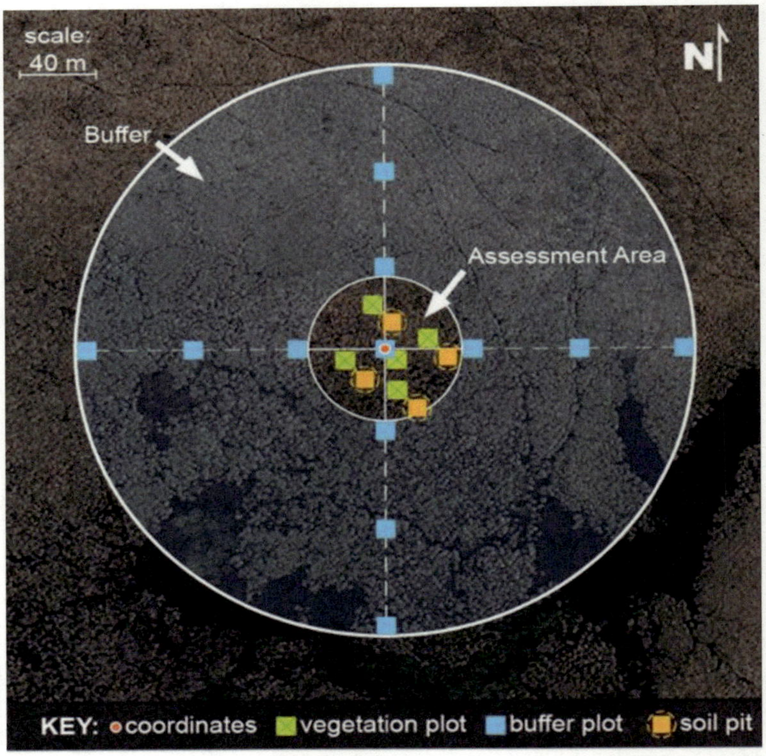

Figure 2.
Diagram of a standard layout for a 0.5-ha assessment area and surrounding 100-m buffer (adapted from USEPA [10]). Locations of the coordinates for the site location generated by the survey design, of vegetation and buffer plots, and of soil pits are indicated.

AA unfeasible, alternate configurations of the AA and buffer were established using a rule-based system [10]. Sample plots were established in the AA and buffer according to standardized protocol to collect observational data and samples associated with physical, chemical, and biological aspects of each site.

Physical aspects of the site were characterized by evidence of human activities in the AA and buffer. Using a standardized checklist of 52 predefined human activities, field crews collected observational data associated with anthropogenic disturbance from thirteen 100-m^2 plots (one in the center of the AA; 12 in the buffer), and on hydrologic alterations throughout the entire AA (**Figure 2**) [10, 21].

Chemical aspects of the site were characterized using nutrient and heavy metal data associated with soil and surface water samples. To collect soil samples, field crews first excavated four soil pits (**Figure 2**), describing each soil horizon to a depth of 60 cm [10]. Crews chose the pit that best reflected the soils on the site based on the descriptions of the soil horizons and expanded it to 125 cm, collecting soil samples for each horizon. Soil samples were analyzed for heavy metals and phosphorus, among other analytes, by the US Department of Agriculture, Natural Resource Conservation Service, Kellogg Soil Survey Laboratory, Lincoln, Nebraska, using standard procedures [11, 22]. Surface water samples were collected as close to the center of the AA as possible at sites where adequate ($\geq$15 cm deep) surface water was present in the AA and prior to conducting other sampling activities to avoid disturbance of the water and substrate, and before 1100 h to avoid diurnal changes in the chemistry [10]. The characteristics of the location from which the water sample was collected were recorded, including the stage of tide for tidal sites.

Biological aspects of the site were characterized using vegetation data. Field crews recorded plant species identity and abundance data in five, systematically placed, 100-m^2 vegetation plots within the AA (**Figure 2**) [10, 11, 23, 24]. A variety of information describing attributes of vegetation structure was also collected within each plot.

3. Understanding what *wetland water chemistry* represents

The value of the probability design used in the NWCA is that the wetland sites sampled represent the larger population of wetlands that meet the target definition. In other words, data that were collected at the 967 wetland sites sampled in 2011 can be inferred to 25,153,681 ha of wetland area across the conterminous US.

Most kinds of data were collected at all wetland sites; however only a portion of the sites had surface water during the 2011 field visits, so water chemistry samples could be collected from just 537 sites of the 967 sampled sites. Factoring in the design weights from the sites with water samples, only 41%, or 10,408,004 ha of the 25,153,681 ha of total sampled population was represented by surface water chemistry. In addition, the 10,408,004 ha represented by water chemistry data do not represent the total sampled population. This is most evident in the proportion of wetland area with water chemistry data in each of the five ecoregions (**Figure 3a**). Surface water chemistry was most commonly sampled in the Tidal Saline (TSL) and Interior Plains (IPL) regions, and represented 72 and 62%, respectively, of the total sampled wetland area. Water chemistry data for the Coastal Plains (CPL), Eastern Mountains and Upper Midwest (EMU), and West (W) represented, respectively, 33, 34, and 47% of the estimated wetland area in each of these ecoregional subpopulations. The proportion of wetlands with surface water in each ecoregion is driven by climatic differences [25], and by characteristics of the landscape [26–29].

Characteristics of the landscape that drive wetland structure and function are embodied in the hydrogeomorphic (HGM) classification [30, 31]. HGM wetland

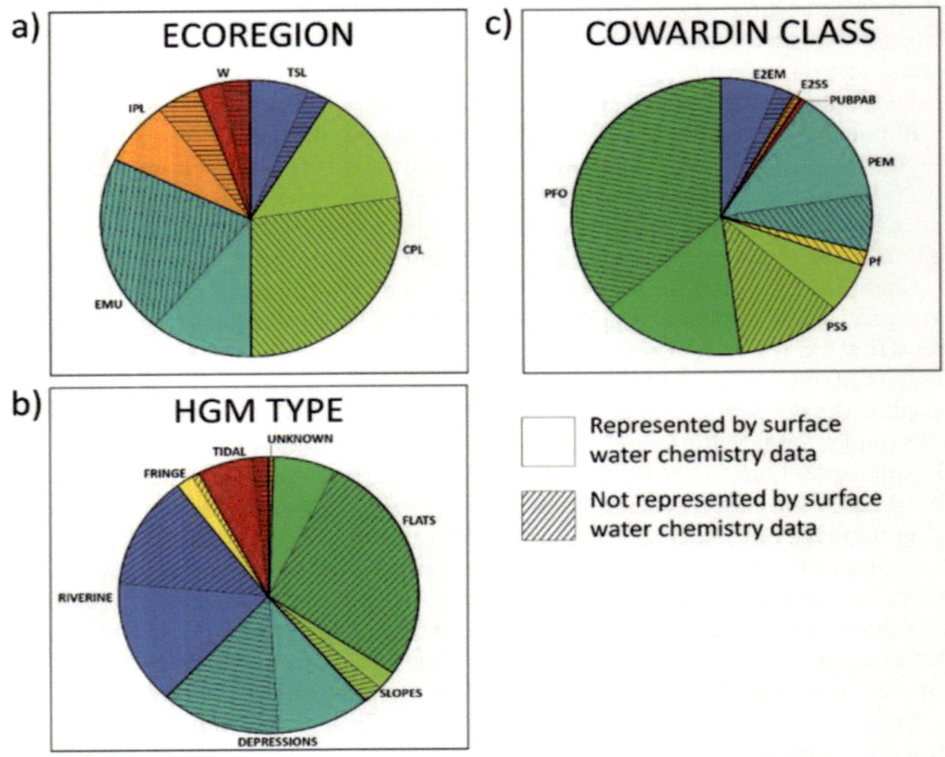

Figure 3.
Proportional area of the 2011 National Wetland condition assessment (NWCA) sampled wetland population represented (solid wedges) and not represented (hatched wedges) by surface water chemistry data. The sampled wetland population is presented using three different wetland groupings: (a) Ecoregion (TSL = Tidal Saline, CPL = Coastal Plains, EMU = Eastern Mountains and Upper Midwest, IPL = Interior Plains, and W = West), (b) hydrogeomorphic (HGM) type, and (c) Cowardin Class (E2EM = Estuarine Intertidal Emergent, E2SS = Estuarine Intertidal Forested/Scrub Shrub, PUBPAB = Palustrine Unconsolidated Bottom/ Aquatic Bed, PEM = Palustrine Emergent, Pf = Palustrine Farmed, PSS = Palustrine Scrub Shrub, and PFO = Palustrine Forested). For HGM type, unknown represents wetland area that was unable to be classified by the field crews. Note that solid and hatched wedges within the same color together represent 100% of the sampled wetland area within the subpopulation.

types are flats, slopes, depressions, riverine, fringe, and tidal [30–33]. These types are arranged along a hydrologic gradient from the least to the most surface water in **Figure 4**. Perhaps, unsurprisingly given that flats have the least surface water, water chemistry data only represented 20% of the total area of flats in the sampled population (**Figure 3b**). Conversely, tidal and fringe HGM types, which tend to have the most surface water throughout the year, had water chemistry data for 77 and 71% of their sampled wetland area, respectively. Slopes, depressions, and riverine wetlands encompass a wide range of varying hydrologic regimes; about half of the wetland area each of these HGM types were represented by water chemistry data (51, 44, and 52%, respectively).

While HGM classifies wetlands based on a hydrologic gradient, Cowardin wetland classes [4], used in the NWCA design, characterizes wetlands by the type of dominant vegetation. Again, the water chemistry does not equally represent the total sampled wetland area associated with each class. Wetland classes dominated by floating and rooted submerged vegetation (PUBPAB) and emergent herbaceous vegetation (E2EM, PEM) are better represented by the water chemistry data than are wetland classes dominated by forest (PFO) and shrub scrub (E2SS, PSS).

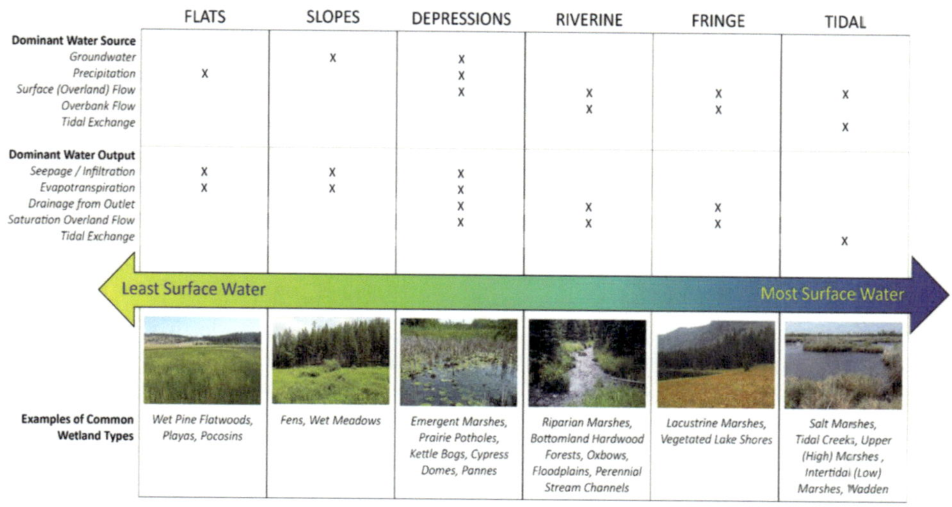

Figure 4.
The gradient of hydrologic conditions associated with hydrogeomorphic (HGM) wetland types as characterized by their dominant water sources and water outputs. Photos exemplifying each HGM wetland type were taken by the NWCA field crews.

Figure 3c shows that 94% of PUBPAB, 76% of E2EM, 66% of PEM wetland area was represented by surface water chemistry data, while only 30% of PFO, 33% of E2SS, and 34% of PSS wetland area were represented by surface water chemistry data.

Our results from the NWCA show that using water chemistry to determine whether the wetland resource meets CWA criteria poses a number of issues. Wetland water chemistry data are biased relative to ecoregions, HGM wetland types, and Cowardin wetland classes [4]. This is because water chemistry data tend to capture wetlands that are permanently flooded, clearly under-representing precipitation- and groundwater-driven wetlands and wetland types that drawdown during the summer (i.e., when sites were sampled). Wetlands dominated by herbaceous vegetation were better, but far from completely, represented by the 2011 water chemistry data, compared to wetlands dominated by woody vegetation where only a third of the area was represented. Water chemistry is often seen as a fundamental component for monitoring and evaluating aquatic systems; however, in the case of the majority of wetlands (where the presence of surface water is highly variable) interpreting what water chemistry results represent and what they signify is problematic.

4. Measuring *wetland resource quality* through the 2011 NWCA

Surface water chemistry as an indicator of chemical integrity is limited to wetland types that have permanent or recurrent surface water or require continuous monitoring throughout the year to capture wetland types that have ephemeral or infrequent surface water. Collecting surface water samples from some wetland types that rarely have surface water, like flats and slopes (**Figure 4**), may be unfeasible. Water chemistry is not a consistently available, readily interpretable, indicator for wetlands across the nation.

Fortunately, there are physical, chemical, and biological indicators of integrity that can be measured consistently and are easily interpreted. Using a suite of physical, chemical, and biological indicators to describe condition also directly addresses the recommendations in the CWA. The 2011 NWCA illustrates how physical, chemical, and biological indicators were employed as the basis for assessing condition.

4.1 Use of condition to report on the state of *wetland resource quality*

Condition of an ecosystem can be expressed in different ways—ecological, or by individual components (biological, chemical, physical). Biological condition of the wetland resource at national and regional scales in the 2011 NWCA was used to report on the state of *wetland resource quality*. To evaluate the biological condition of wetlands, a multimetric index was developed based on plant species and trait data collected as part of the NWCA [12, 23]. Although the Vegetation Multimetric Index (VMMI) is biological in nature, it is calibrated using physical, chemical, and biological data that reflect the level of anthropogenic disturbance at a site.

Physical, chemical, and biological data resulting from information collected in the field were used to construct 10 measures of anthropogenic disturbance [10, 34]. Eight indices utilized observational data to describe physical disturbance [21]; one index used concentrations of heavy metals in the wetland soils to describe chemical disturbance [22]; and one metric for relative cover of alien plant species was used to describe biological disturbance. For each of the 10 measures, thresholds were established to reflect the degree of human impact to the site. A screening approach was used to categorize sites as least disturbed, moderately disturbed, or most disturbed based on the frequency at which thresholds were exceeded [12, 34]. Least disturbed sites, which represented the best attainable conditions given the state of the landscape [35], were used as a measure of physical, chemical, and biological reference condition in developing the VMMI.

Development of the VMMI is described in detail in Magee et al. [23] and began with calculation of 405 candidate metrics describing different vegetation properties with probable relationships to biological condition. The potential efficacy of each metric in reflecting biological condition was evaluated using a variety of objective screening tests with cut-offs appropriate to wetland data including: (1) sufficient range in values to allow detection of signals in response to disturbance; (2) repeatability, quantified using a signal to noise ratio (S:N) based on repeat sampling of a subset of sites (see Magee et al. [23] for a discussion of S:N); and (3) responsiveness, that is, how well a metric distinguished least disturbed from most disturbed wetland sites sampled in the NWCA. Candidate metrics that passed the screening criteria were examined for utility as components of potential VMMIs. Many thousands of potential VMMIs combining from 4 to 10 individual metrics were calculated and evaluated using approaches similar to Van Sickle [36] and Stoddard et al. [37], but adapted for wetlands, to identify the VMMIs with the best performance and with limited redundancy (correlation) among metrics included in a particular VMMI [23]. The final national-scale VMMI for the 2011 NWCA was based on the combination of four metrics, all broadly applicable across major classes of wetlands (**Table 1**). The VMMI is scaled from 0 to 100, with higher values representing better biological condition. To translate the continuous VMMI scores to condition categories, thresholds for delineating "good," "fair," and "poor" condition were determined based on the distribution of VMMI values in least-disturbed sites [23] using the percentile approach described in Paulsen et al. [38].

Biological condition of wetlands, reported as "good," "fair," and "poor" by the 2011 NWCA, reflects the state of the *wetland resource quality* as measured at all 967 sampled wetland probability sites, representing 25,153,681 ha of wetlands across

Metric name	Metric description	Calculation
Floristic Quality Assessment Index (FQAI)	Based on all species observed	**FQAI = $\Sigma CC_{ij}/\sqrt{N_j}$** where CC_{ij} = coefficient of conservatism for each unique species *i* at site *j* N = number of species at site *j*
Relative importance of native species	Combines relative cover and relative frequency for native taxa at each site	**((Σ Absolute Cover native species$_i$/Σ Absolute Cover all species$_i$) × 100 + (Σ Frequency native species$_i$/Σ Frequency all species$_i$) × 100)/2** where for each unique species *i*: absolute cover = 0–100%, frequency = 0–100% calculated as the percent of plots in which it occurred
Richness of disturbance-tolerant species	Tolerance to disturbance defined as coefficient of conservatism (C-value) ≤ 4	**Number of taxa with C-value ≤ 4 occurring at a site**
Relative cover of native monocots	Relative cover of native monocot species at each site	**(Σ Absolute Cover native monocot species$_i$/Σ Absolute Cover all species$_i$) × 100**

Table 1.
The four metrics, and equations for their calculation at each sampled site, that were included in the 2011 National Wetland Condition Assessment (NWCA) vegetation multimetric index (VMMI) as described in Magee et al. [23].

the conterminous US. Specifically, results from the survey showed that 48% of the target sampled wetland area in the nation was in good condition, 20% was in fair condition, and 32% was in poor condition (**Figure 5**) [8].

4.2 Evaluation of *wetland resource quality* using indicators of stress

While condition describes the state of *wetland resource quality*, it is equally important to understand factors that negatively affect *wetland resource quality* in making policy and resource management decisions. This requires an evaluation using physical, chemical, and biological stressor data [3]. The concepts of relative extent and relative risk were used to report the magnitude of six physical indicators of stress [21], two chemical indicators of stress [22], and one biological indicator of stress across wetlands of the US [24] to evaluate the impact of the chemical and physical stressors on the state of the *wetland resource quality* [39].

Using observational data collected in the buffer and in the assessment area (AA), an Anthropogenic Stress Index (ASI) was developed for six physical stressor categories: vegetation removal, vegetation replacement, damming, ditching, hardening, and filling/erosion (**Table 2**). Thresholds that indicate the degree of physical stress associated with each physical stressor were established [12, 21]. Each site was assigned to either low, moderate, or high stressor levels for each of the six stressor categories based on its ASI score.

Soil chemistry data were examined to identify chemical indicators of stress. Ultimately, only heavy metals and total phosphorus concentrations were used in the NWCA analysis (**Table 2**). Twelve heavy metals, each (1) with high signal-to-noise ratios [40], (2) a close relation to anthropogenic impacts, and (3) occurring in consistently measurable quantities, were used to develop a Heavy Metals Index (HMI) [12, 22]. The metals were: silver, cadmium, cobalt, chromium, copper, nickel, lead, antimony, tin, vanadium, tungsten, and zinc. The HMI is the sum of the number of metals present in the uppermost layer of the soil with concentrations above expected natural background levels. Background levels were based on published values primarily from Alloway [41] and used directly or slightly modified.

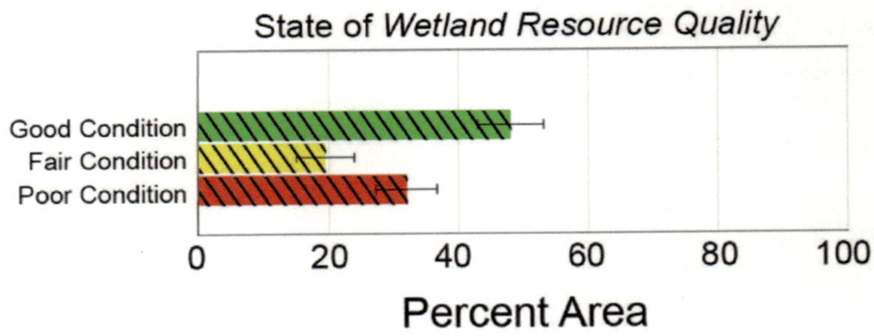

Figure 5.
State of the wetland resource quality *as indicated by the vegetation multimetric index (VMMI) for the 2011 National Wetland Condition Assessment (NWCA). Condition classes are reported as the percent area of the sampled wetland population. Error bars are 95% confidence intervals (figure adapted from USEPA [8]).*

Because no published thresholds for anthropogenic impacts to wetlands were available, thresholds for chemical stressor levels were set based on the background concentrations from Alloway [12, 22, 41]. The threshold for the low HMI stressor level required that all metals were less than or equal to background concentrations, and the threshold for the high HMI stressor levels was ≥3 metals above background. All values falling between the high and low stressor levels were termed moderate. In the case of phosphorus, concentration of total phosphorus in the uppermost layer with soil chemistry was used as a chemical indicator of stress. The thresholds for low and high phosphorus stressor levels were set using the 75th and 95th percentiles observed in least-disturbed sites [42, 43].

The Nonnative Plant Indicator (NNPI) was developed as a biological indicator of stress [12, 24]. Nonnative plants are widely recognized as (1) indicators of stress (e.g., their presence is often associated with human-mediated disturbances that negatively affect biological condition), or as (2) direct stressors to the condition of wetlands and other ecosystems (e.g., by inducing structural changes in vegetation, competing with native plant species, altering species interactions, community composition, or ecosystem properties); see Magee et al. [24] and citations therein. The NNPI is a categorical indicator based on three metrics describing different pathways of potential effects from the collective set of nonnative taxa occurring at each site (**Table 2**). The three NNPI metrics (nonnative relative cover, nonnative richness, and nonnative relative frequency) were used together in a decision matrix to assign each sampled site to a stressor-level category (low, moderate, or high) based on exceedance values for each metric [12, 24]. Note, that the high stressor-level category presented here combines the high and very-high stressor levels defined in Magee et al. [24].

Relative extent describes the frequency at which indicators of stress occur in wetlands and can be used to identify the most common indicators of stress occurring at high levels likely affecting wetland resource quality. Using the low, moderate, and high stressor-level thresholds for each of the indicators of stress, the wetland area associated with each stressor level and indicator was determined using the weights from the sampled sites [39]. Relative extent is reported as the proportion of wetland area sampled with high stressor levels for each of the indicators of stress (**Figure 6**). The most frequently encountered indicators of stress at high stressor levels were associated with physical indicators and include vegetation removal, hardening, and ditching, at 27, 27, and 23% of the sampled wetland area, respectively. The NNPI had 19% area associated with the high stressor level, while the chemical indicators, soil phosphorus, and heavy metals, had 6 and 2% of the sampled wetland area associated with the high stressor level.

Relative risk can be used to evaluate the proportional effect of factors that have an impact on wetland resource quality and is defined as the probability of having

Indicators	Description	Observations/measurements included
Physical indicators		
Vegetation removal	Any field observation related to loss, removal, or damage of wetland vegetation	*Gravel pit, oil drilling, gas wells, underground mine, forest clear cut, forest selective cut, tree canopy herbivory, shrub layer browsed, highly grazed grasses, recently burned forest, recently burned grassland, herbicide use, mowing/shrub cutting, pasture/hay, range*
Vegetation replacement	Any field observation of altered vegetation within the site due to anthropogenic activities	*Golf course, lawn/park, row crops in small amounts in the Assessment Area, row crops in the buffer, fallow field, nursery, orchard, tree plantation*
Damming	Any field observation related to impounding or impeding water flow from or within the site	*Dike/dam/road/RR bed, water level control structure, wall/riprap, dikes, berms, dams, railroad beds, sewer outfalls*
Ditching	Any field observation related to draining water	*Ditches, channelization, inlets/outlets, point source/pipe, irrigation, water supply, field tiling, standpipe outflow, corrugated pipe, box culvert, outflowing ditches*
Hardening	Any field observation related to soil compaction, including activities and infrastructure that primarily result in soil hardening	*Gravel road, two-lane road, four-lane road, parking lot/pavement, trails, soil compaction, off road vehicle damage, confined animal feeding, dairy, suburban residential, urban/multifamily, rural residential, impervious surface input, animal trampling, vehicle ruts, roads, concrete, asphalt*
Filling/erosion	Any field observation related to soil erosion or deposition	*Excavation/dredging, fill/spoil banks, freshly deposited sediment, soil loss/root exposure, soil erosion, irrigation, landfill, dumping, surface mine, recent sedimentation, excavation/dredging*
Chemical indicators		
Heavy Metal Index	Heavy metals with concentrations above background concentrations in soil samples	*Antimony, cadmium, chromium, cobalt, copper, lead, nickel, silver, tin, tungsten, vanadium, zinc concentrations from the uppermost layer with soil chemistry*
Soil phosphorus concentration	Soil phosphorus concentrations relative to reference sites	*Phosphorus concentration from the uppermost layer within 10 cm of the soil surface with soil chemistry*
Biological indicator		
Nonnative Plant Indicator (NNPI)	A categorical indicator based on three metrics that describe different avenues of potential impact to biological condition	*Relative cover of nonnative species, richness of nonnative species, relative frequency of nonnative species*

Table 2.
Description and components of the biological, physical, and chemical indicators of stress (adapted from USEPA [12]).

poor condition when stressor levels are high relative to when stressor levels are low [12, 39, 44–46]. Relative risk was calculated for the six physical and two chemical indicators of stress. Because condition of wetlands is based on vegetation data (i.e., the VMMI) and the biological indicator of stress (i.e., the NNPI) also uses the vegetation data, relative risk is not reported for the NNPI (see [12] for details). **Figure 6** shows the relative risk for the physical and chemical stressors. The likelihood of poor condition (compared to good condition) was 1.8 times higher when

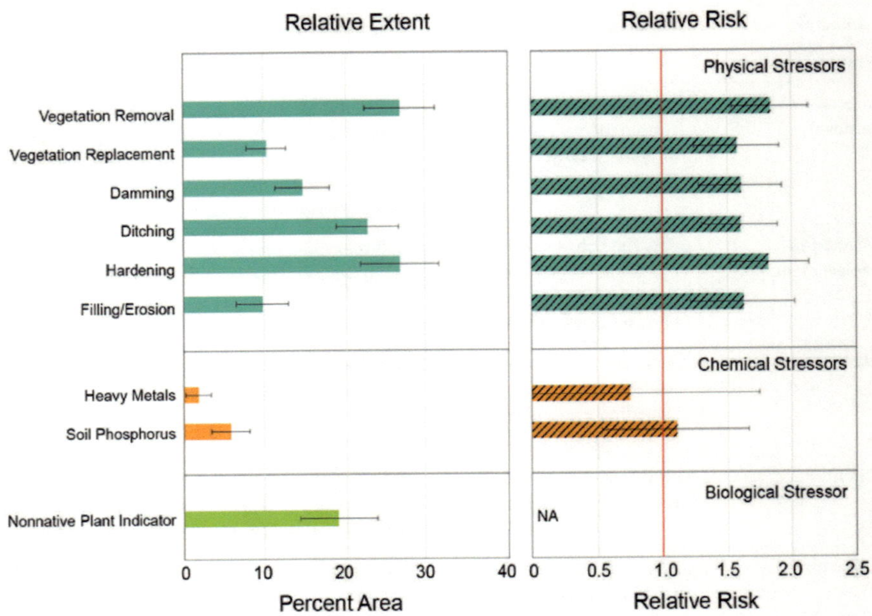

Figure 6.
Evaluation of the factors that affect wetland resource quality as indicated by relative extent (percent area of the wetland resource) and relative risk from chemical, physical, and biological indicators of stress for the 2011 National Wetland Condition Assessment (NWCA). NA indicates "not applicable" for relative risk of the nonnative plan indicator to avoid circularity (see text for details). Error bars are 95% confidence intervals (figure adapted from USEPA [8]).

vegetation removal and hardening are present at high stressor levels and 1.6 times higher when vegetation replacement, damming, ditching, and filling/erosion are present at high stressor levels. A relative risk of 1.0 indicates that there is no association or relationship between the indicator of stress and condition, and a relative risk less than 1.0, indicates a positive relationship between high stressor level of the indicator and good condition.

4.3 Summary of *wetland resource quality* in the conterminous US

The results of the 2011 NWCA indicates that the *wetland resource quality* across the conterminous US is good for about half of the wetland area, with the remainder divided between fair and poor *wetland resource quality* (**Figure 5**). Physical, chemical, and biological data collected in the field can also be used to evaluate factors that impact wetland resource quality. Review of the patterns in relative extent of the examined indicators of stress that were found at high stressor level, shows that specific physical stressors and the biological stressor were the most frequently encountered and may affect wetland resource quality, while chemical indicators of stress are less common at high stressor levels (**Figure 6**). The effect of stressors on *wetland resource quality* is illustrated by the relative risk results (**Figure 6**), which show that physical indicators of stress occurring at high stressor levels are likely to impact *wetland resource quality*.

5. Conclusions

We use the NWCA as an example of how physical, chemical, and biological data collected in the field can be synthesized to evaluate the state of *wetland resource quality* (a specific type of *aquatic resource quality*). Furthermore, we illustrate that we can evaluate the factors affecting *wetland resource quality* on a national scale using relative extent and relative risk.

We believe that the concept of *aquatic resource quality* should be the basis for monitoring aquatic ecosystems. First, *aquatic resource quality* reflects condition, which is founded in physical, chemical, and biological data. Therefore, *aquatic resource quality* directly addresses the CWA goals for reporting on the physical, chemical, and biological integrity of water resources. Secondly, the concept of *aquatic resource quality* can be evaluated in all aquatic ecosystems, regardless of surface water availability (as in the case of precipitation-driven wetlands and ephemeral streams) or aquatic ecosystem type (e.g., wetlands versus streams). In fact, the data needed to evaluate *aquatic resource quality* in all aquatic ecosystems across the conterminous US are already being collected through the Environmental Protection Agency's (USEPA) National Aquatic Resource Surveys (NARS) program (https://www.epa.gov/national-aquatic-resource-surveys). Physical, chemical, and biological data are collected every year from one of four aquatic resources—rivers and streams, lakes, wetlands, and coasts—to assess the status of their condition. The NWCA, discussed extensively here, is the wetland component of NARS. Every 5 years, the entire water resource of the nation is assessed, allowing appraisal of trends and changes over time. In addition, because condition, relative risk, and relative extent are measured using comparable design, field protocols, and analysis methods for all aquatic resources assessed in NARS, the opportunity exists to evaluate and compare *aquatic resource quality* across ecosystem types on a national scale.

Another advantage of adopting *aquatic resource quality* as the basis for monitoring aquatic ecosystems is that the results are easily translatable to a non-scientific audience, in part because the concepts and terminology are unambiguous. In addition, information surrounding *aquatic resource quality* can be reported in a way that answers questions of interest to the public. These might include:

- What is the state of the *aquatic resource quality*?
- What factors are negatively affecting *aquatic resource quality*?
- How do the patterns in *aquatic resource quality* change over time?

Moreover, the questions can be addressed using tested and established NARS methods to gather data for reporting on *aquatic resource quality* using condition, relative extent, and relative risk. NARS field, laboratory, and analysis methods are publicly available and applicable to multiple scales. NARS methodology allows for the consideration of results beyond the context of individual sampled sites, thus increasing the power of the data. For example, (a) results can be compared to regional and national NARS datasets, or (b) results can be compared or combined with those from other data collected using the NARS methodology.

The example in this chapter was national in scale and evaluated wetlands; however, sampling physical, chemical, and biological indicators to characterize condition can also be applied to regional, state, and local aquatic ecosystems. *Aquatic resource quality* is broadly relevant and supports management and policy decisions across ecosystem types, spatial scales, and political entities.

Acknowledgements

The NWCA was planned, funded, and organized by the US Environmental Protection Agency's (USEPA) Office of Water (OW) and Office of Research and

Development (ORD), and implemented by numerous state, federal, and contractor field crews, information management staff, and laboratory staff whose efforts the authors gratefully acknowledge. The authors appreciate the thoughtful reviews provided by Alan Herlihy, Oregon State University, Gregg Serenbetz, USEPA OW, and James Markwiese, USEPA ORD. Their input greatly improved the manuscript.

This chapter has been subjected to the USEPA's review process and has been approved for publication. The views expressed in this chapter are those of the authors and do not necessarily reflect the views or policies of the agency. Any mention of trade names, products, or services does not imply an endorsement by the US Government or the USEPA. The USEPA does not endorse any commercial products, services, or enterprises. This work was authored by US Government employees as part of their official duties. In view of Section 105 of the Copyright Act (17 U.S.C. §105), this chapter is not subject to US copyright protection.

Author details

Mary E. Kentula*, Amanda M. Nahlik, Steven G. Paulsen and Teresa K. Magee
US Environmental Protection Agency, Office of Research and Development, Center for Public Health and Environmental Assessment, Pacific Ecological Systems Division, Corvallis, Oregon, USA

*Address all correspondence to: kentula.mary@epa.gov

References

[1] Eriksson E. Water chemistry and water quality. Ambio. 1977;**6** (1, Water, A Special Issue):27-30

[2] Federal Water Pollution Control Amendments of 1972 (L. 92-500). Sect. 33 U.S.C. Section 1251 et seq.; 1972

[3] Paulsen SG, Hughes RM, Larsen DP. Critical elements in describing and understanding our nation's aquatic resources. Journal of the American Water Resources Association. 1998;**34**(5):995-1005

[4] Cowardin LM, Carter V, Golet FC, LaRoe ET. Classification of Wetlands and Deepwater Habitats of the United States. Washington, DC: U.S. Fish and Wildlife Service; 1979. Report No.: FWS/OBS-79/31

[5] Wardrop DH, Kentula ME, Stevens DL Jr, Jensen SF, Brooks RP. Assessment of wetland condition: An example from the upper Juniata watershed in Pennsylvania, USA. Wetlands. 2007;**27**:416-430

[6] Whigham DF, Deller Jacobs A, Weller DE, Jordan TE, Kentula ME, Jensen SF, et al. Combining HGM and EMAP procedures to assess wetlands at the watershed scale - status of flats and non-tidal riverine wetlands in the Nanticoke River watershed, Delaware and Maryland (USA). Wetlands. 2007;**27**(3):462-478

[7] Brooks RP, Wardrop DH, editors. Mid-Atlantic Freshwater Wetlands: Advances in Wetlands Science, Management, Policy, and Practice. New York, NY: Springer; 2013

[8] USEPA. National Wetland Condition Assessment 2011: A Collaborative Survey of the Nation's Wetlands. Washington, DC: U.S. Environmental Protection Agency; 2016. Report No.: EPA/843/R-15/005 Contract No.: EPA/843/R-15/005

[9] USEPA. National Wetland Condition Assessment 2011: Site Evaluation Guidelines. Washington, DC: U.S. Environmental Protection Agency; 2011. Report No.: EPA/843/R-10/004

[10] USEPA. National Wetland Condition Assessment 2011: Field Operations Manual. Washington, DC: U.S. Environmental Protection Agency; 2011. Report No.: EPA/843/R-10/001

[11] USEPA. National Wetland Condition Assessment 2011: Laboratory Methods Manual. Washington, DC: U. S Environmental Protection Agency; 2011. Report No.: EPA/843/R-10/002

[12] USEPA. National Wetland Condition Assessment 2011: Technical Report. Washington, DC: U.S. Environmental Protection Agency; 2016. Report No.: EPA/843/R-15/006 Contract No.: EPA/843/R-15/006

[13] Dahl TE, Bergeson MT. Technical Procedures for Conducting Status and Trends of the Nation's Wetlands. U.S. Fish and Wildlife Service, Division of Habitat and Resource Conservation: Washington, D.C; 2009

[14] Stevens DL, Olsen AR. Spatially restricted surveys over time for aquatic resources. Journal of Agricultural, Biological, and Environmental Statistics. 1999;**4**:415-428

[15] Stevens DL, Olsen AR. Spatially restricted random sampling designs for design-based and model-based estimation. In: Accuracy 2000: Proceedings of the 4th International Symposium on Spatial Accuracy Assessment in Natural Resources and Environmental Sciences. The Netherlands: Delft University Press; 2000

[16] Stevens DL, Olsen AR. Spatially-balanced sampling of natural resources.

Journal of the American Statistical Association. 2004;**99**(465):262-278

[17] Olsen AR, Kincaid TM, Kentula ME, Weber MH. Survey design to assess condition of wetlands in the United States. Environmental Monitoring and Assessment. 2019;**191**:16

[18] Diaz-Ramos S, Stevens DL Jr, Olsen AR. EMAP Statistics Methods Manual. Corvallis, OR: US Environmental Protection Agency, Office of Research and Development, National Health and Environmental Effects Research Laboratory; 1996. Report No.: EPA/620/R-96/002

[19] Kincaid TM, Olsen AR. Spsurvey: Spatial Survey Design and Analysis: R Package Version 3.1; 2015. Available from: //cran.fhcrc.org/

[20] McCauley DJ, Arnold WJ, Saxton JB, Turner CJ. Applying adaptive management and lessons learned from national assessments to address logistical challenges in the National Wetland Condition Assessment. Environmental Monitoring and Assessment. 2019;**191**:14

[21] Lomnicky GA, Herlihy AT, Kaufmann PR. Quantifying the extent of human disturbance activities and anthropogenic stressors in wetlands across the conterminous United States: Results from the National Wetland Condition Assessment. Environmental Monitoring and Assessment. 2019;**191**:23

[22] Nahlik AM, Blocksom KA, Herlihy AT, Kentula ME, Magee TK, Paulsen SG. Use of national-scale data to examine human-mediated additions of heavy metals to wetland soils of the US. Environmental Monitoring and Assessment. 2019;**191**:24

[23] Magee TK, Blocksom KA, Fennessy MS. A national-scale vegetation multimetric index (VMMI) as an indicator of wetland condition across the conterminous United States. Environmental Monitoring and Assessment. 2019;**191**:28

[24] Magee TK, Blocksom KA, Herlihy AT, Nahlik AM. Characterizing nonnative plants in wetlands across the conterminous United States. Environmental Monitoring and Assessment. 2019;**191**:32

[25] Semeniuk CA, Semeniuk V. The response of basin wetlands to climate changes: A review of case studies from the swan coastal plain, South-Western Australia. Hydrobiologia. 2013;**708**(1):45-67

[26] Winter TC. A physiographic and climatic framework for hydrologic studies of wetlands. In: Robarts RD, Bothwell ML, editors. Aquatic Ecosystems in Semi-Arid Regions: Implications for Resource Management. Saskatoon, Saskatchewan, Canada: Environment Canada; 1992. pp. 127-148

[27] Winter TC. The concept of hydrologic landscapes. Journal of the American Water Resources Association. 2001;**37**(2):335-349

[28] Bedford BL. The need to define hydrologic equivalence at the landscape scale for freshwater wetland mitigation. Ecological Applications. 1996;**6**(1):57-68

[29] Wolock DM, Winter TC, McMahon G. Delineation and evaluation of hydrologic-landscape regions in the United States using geographic information system tools and multivariate statistical analyses. Environmental Management. 2004;**34**(Suppl 1):S71-S88

[30] Brinson MM. A Hydrogeomorphic Classification for Wetlands. Vicksburg, MS: U.S. Army Corps of Engineers,

Waterways Experiment Station; 1993. Report No.: Technical Report WRP-DE-4

[31] Smith RD, Ammann A, Bartlodus C, Brinson MM. An Approach for Assessing Wetland Functions Using Hydrogeomorphic Classification, Reference Wetlands, and Functional Indices. Vicksburg, MS, USA: U.S. Army Corps of Engineers, Waterways Experiment Station; 1995. Report No.: Technical Report WRP-DE-9

[32] Gosselink JG, Turner RE. The role of hydrology in freshwater wetland ecosystems. In: Good RE, Whigham DF, Simpson RL, editors. Freshwater Wetlands: Ecological Processes and Management Potential. New York, NY: Academic Press; 1978

[33] Mitsch WJ, Gosselink JG. Wetlands. 3rd ed. New York, NY: John Wiley; 2000

[34] Herlihy AT, Kentula ME, Magee TK, Lomnicky GA, Nahlik AM, Serenbetz G. Striving for consistency in the National Wetland Condition Assessment: Developing a reference condition approach for assessing wetlands at a continental scale. Environmental Monitoring and Assessment. 2019;**191**:20

[35] Stoddard JL, Larsen DP, Hawkins CP, Johnson PK, Norris RH. Setting expectations for the ecological condition of streams: The concept of reference condition. Ecological Applications. 2006;**16**(4):1267-1276

[36] Van Sickle J. Correlated metrics yield multimetric indices with inferior performance. Transactions of the American Fisheries Society. 2010;**139**:1802-1817

[37] Stoddard JL, Herlihy AT, Peck DV, Hughes RM, Whittier TR, Tarquinio E. A process for creating multimetric indices for large-scale aquatic surveys. Journal of the North American Bethological Society. 2008;**27**(4):878-891

[38] Paulsen SG, Mayio A, Peck DV, Stoddard JL, Tarquinio E, Holdsworth SM, et al. Condition of stream ecosystems in the US: An overview of the first national assessment. Journal of the North American Bethological Society. 2008;**27**(4):812-821

[39] Herlihy AT, Paulsen SG, Kentula ME, Magee TK, Nahlik AM, Lomnicky GA. Assessing the relative and attributable risk of stressors to wetland condition across the conterminous United States. Environmental Monitoring and Assessment. 2019;**191**:17

[40] Kaufmann PR, Hughes RM, Van Sickle J, Whittier TR, Seeliger CW, Paulsen SG. Lakeshore and littoral physical habitat structure: A field survey method and its precision. Lake and Reservoir Management. 2014;**30**(2):157-176

[41] Alloway BJ, editor. Heavy Metals in Soils: Trace Metals and Metalloids in Soils and their Bioavailabilaity. New York, NY: Springer; 2013

[42] Herlihy AT, Paulsen SG, Van Sickle J, Stoddard JL, Hawkins CP, Yuan LL. Striving for consistency in a national assessment: The challenges of applying a reference-condition approach at a continental scale. Journal of the North American Benthological Society. 2008;**27**(4):860-877

[43] Herlihy AT, Sobota JB, McDonnell TC, Sullivan TJ, Lehmann S, Tarquinio E. An a priori process for selecting candidate reference lakes for a national survey. Freshwater Science. 2013;**32**(2):385-396

[44] van Sickle J, Stoddard JL, Paulsen SG, Olsen AR. Using relative

risk to compare the effects of aquatic stressors at a regional scale. Environmental Management. 2006;**38**(6):1802-1817

[45] Van Sickle J, Paulsen SG. Assessing the attributable risks, relative risks, and regional extents of aquatic stressors. Journal of the North American Benthological Society. 2008;**27**(4):920-931

[46] Van Sickle J. Estimating the risks of multiple, covarying stressors in the National Lakes Assessment. Freshwater Science. 2013;**32**(1):204-216